Asim Kumar Mukhopadhyay

Essential Oils

Natural and Synthetic

Amani International Publishers

Bibliographic information published by the Deutsche Nationalbibliothek

The Deutsche Nationalbibliothek lists this publication in the Deutsche Nationalbibliografie; detailed bibliographic data are available in the Internet at http://dnb.d-nb.de

Manufactured in Germany

E-Mail: service@amani-publishers.com
http://www.amani-publishers.com

ISBN: 978-3-938054-32-1
eISBN: 978-3-938054-38-3

Dedicated to my mother whose never ending love for fragrances emanating from flowers like rose and jasmine and flavors based on clove and cardamom,
prompted me to write this book on the subject.

Overview

Contents

Preface

My childhood and boyhood days and years were spent in a sleepy hamlet in the district of Bankura, West Bengal, India. In the midst of Serenity and tranquility of the natural environment a large number of flowering plants used to be born either as wild growths or cultivated. Some of the more popular flowers were rose, jasmine, marigold, lotus, water lilies etc. It is thus no wonder that I developed a natural liking for these fragrant flowers and other parts of such odoriferous plants. Agarbatis or perfumed essence sticks traditionally used for puja's or worshipping or gods further acted as catalysts for my immense love of flavors and fragrances in general and main constituents e.g,, essential oils or volatile oils, in particular.

Essential oils are the odorous constituents in various plant parts. At ordinary or ambient temperatures they evaporate easily when exposed to air and hence they are also called volatile oils. They are called essential oils because they represent the essence or active constituents of the plants. Essential oils are found in buds, flowers, leaves, bark, stems, fruits, seeds, wood, roots and other parts of fragrant plants and herbs etc.

Essential or volatile oils, also known as aromatic oils have been used for hundreds of years. These oils either as individual constituents recovered from aromatic plants or as a mixture of a few essential oils from one or more sources, as a perfume was known in the ancient civilizations of Egypt, India, China, Greece, Rome and other countries. These were all as old as 4000 - 4500 B.C. In other words known ancient civilizations were generous in use of perfumes created from natural herbs and trees. Mother nature was indeed very liberal and generous in creation of a number of aromatic essential oils in botanicals for the benefit of humanity.

Modern scientists, blessed with various tools, not only succeeded in isolation of essential oils from natural plants but were also successful in synthesis of various essential oils or components of flavors and fragrances. These synthetic oils or flavors and fragrances being relatively cheap compared to natural products have been liberally used for creation of synthetic perfumes and in diversified fields encompassing medicine, organic chemical synthesis, cosmetics, toilet goods, incense sticks, and others.

Many associations and societies active in the fields of flavors and fragrances have been operating in various countries. For instance, Flavors and Fragrances Association of India (FAFAI) has been organizing a number of Seminars and Conferences and also publishing proceedings of annual conferences, and , and books and magazines which are extremely useful for continued research and development work in many of the known universities, national R&D Centers, Industrial laboratories etc. India has already established itself as a major producer of such important essential oils as Menthol, Lemon grass oil, Rose oil, Jasmine oil, Sandalwood oil, Eucalyptus oil etc. FAFAI has played a key role in popularizing essential oils including the above oils.

THE AUTHOR

Dr. Asim Kumar Mukhopadhyay, the author of this book is a techno-commercial professional having had his earlier education in the district town of Bankura, W. Bengal, India. He graduated in Chemical Engineering from Jadavpur University, Calcutta and earned his Master's degree (M.Tech.) in Chemical Engineering from Indian Institute of Technology, Kharagpur, W. Bengal. He received his Ph.D. in Chemical Technology from Moscow Institute of Petrochemicals and Gas Industries, Moscow, the then Capital of U.S.S.R.

After completion of his studies, Dr. Mukhopadhyay joined the industry and developed himself as a multi-disciplinary professional by handling different portfolios in diversified disciplines of Project Management, R&D Management, Production and Marketing Management, Financial Management etc.

Dr. Mukhopadhyay held top management positions in organizations of repute, such as Managing Director of WEBCON, a multifunctional development consultancy organization promoted by Industrial Development Bank of India and other financial and promotional Institutions and Banks, a whole-time Director of IVP Limited, Mumbai and Chief Executive Officer of Atul Limited, Valsad, Gujarat State, India in its Aromatics Division.

For the last twenty years Dr. Mukhopadhyay was actively associated with development and production of Fine Chemicals including Essential Oils, Flavors and Fragrances. He was not only interested in natural products but also in organic chemical synthesis of quite a few fine chemicals including flavor and fragrance compounds.

A widely traveled person, Dr. Mukhopadhyay is a versatile writer having authored several technical articles, short stories, poems and four books. He is truly a multi-dimensional personality.

Acknowledgement

It is with great pleasure that I record a sense of appreciation to various authors, publishers and professionals who had assisted me directly or indirectly in providing vital information and data for completion of the manuscript on this popular subject.

I take this opportunity to express my deep sense of gratitude to the Librarian, Assistant Librarian and their colleagues of the Indian Institute of Technology, Bombay, India, for using their library facilities.

I must express my sincere appreciation and gratitude to the vastly qualified and knowledgeable professionals in India and abroad who helped me in my global research on the related business.

1

INTRODUCTION

1.1 Backdrop

More than seven thousand years ago in the ancient civilizations of Egypt, India, China and other countries prolific use of flavor and fragrance compounds, and more particularly, essential or volatile aromatic oils was reported. Natural perfumes of plant origin either as pure individual oils or as a blend of a few perfumed oils from two or three different plants or herbs were quite popular. Queen Nefertiti of Egypt, and many years later, queen Cleopatra also of Egypt, made intensive uses of fragrances/perfumes to make themselves more attractive and appealing to their favorite kings, heroes, and other personalities.

In ancient Indian Vedic literature was mentioned the uses of essential oils for various purposes. Ancient saints and sages believed in plain living and high thinking and definitely knew the uses of perfumes or pleasant fragrances of plant origin. Many of such oils were used in aroma therapy and during worshipping of gods and goddesses. Scented Agarbattis/ Dhups etc. (Incense Sticks) were used in religious ceremonies. Medicinal uses of many of these oils were also known to the ancient Rishis or Moonies (Saints). Offerings were made to Firegod or Agni in rituals and religious rites. No puja or worshipping of gods was complete without cleansing of the environment with burnt incense sticks, fruits and fragrant flowers.

The Holy Gita, the sacred book of the hindus, mentions Celestial perfumes to adorn the mightiest of the mightly Lord Krishna.*, ** Many celestial ladies mentioned in ancient Indian literature (Vedas) such as

* The Gita Chapter 11, Text 11:

divya-malyambara-dharam
divya-gandhamulepanam
Sarascarya-mayam devam
Anantam visvato-mukham

Menaka, Urvashi, Rati and even goddess Parvati were reported to have used fragrant flowers of Jasmine, Mogra, Champa, Sandal Wood etc. to enhance their appearance and charm despite having unique and unparallel celestial beauty.

This ancient appreciation of perfumes and fragrances either in the form of whole flowers, leaves or as isolated oils have also found references in Greek mythology and Chinese literature.

Fragrant Substances (odorants) in the form of incense and perfumed unguents were used for ceremonial purposes in ancient times among the Greeks and Romans as luxury items. It is however during the period of Renaissance that isolation of essential oils from green plants, bark, wood, herbs and shrubs and multiple uses of fragrances and perfumery items were explored.

Eau de Cologne, a world famous product had it origin in the city of Cologne or Köln in Germany around 1700 A.D. The classic formula for this product consists of bergamot oil, lemon oil, orange oils and orange flower oil in pure ethanol.

Soon it was found that natural fragrances from plant sources and animals were inadequate and scientists started working on synthesis of flavors and fragrance components, i.e., essential oils. Chemical synthesis of a large number of fragrances and flavors laid the foundation of innovations that eventually resulted in development of art and science of perfumes production and uses in cosmetics, toiletries, soaps and household or industrial products. Soon the subject of "aroma therapy" and medicinal applications of essential oils was developed and perfected for commercial uses for one and all.

1.2 Current Trends

From the ancient civilizations to modern age when science and technology has made very impressive progress and has made marks in all

** The universal form of all-pervading God, Lord Krishna was decorated with many celestial ornaments and bore many divine upraised weapons. He wore celestial garlands and garments. Many divine scents were smeared over his body. All was wonderful, brilliant, unlimited, all expanding.

spheres of life it is but natural that flavors and fragrances industry is no exception, and has made very remarkable growth both qualitatively and quantitatively. Nature crated plants and herbs and shrubs containing essential oils for creation of flavors and fragrances. Many processes particularly extraction and distillation enabled man to isolate pure essential oil components from natural sources.

Through ages of scientific research men learned the techniques of analysis and were more or less successful in finding out the ingredients that these natural products consisted of. Flavors and fragrances from natural sources were not adequate to meet the total demand and it was necessary to bridge the demand-supply gap through synthetically made products. Synthetic essential oils or perfumery chemicals are also cheaper than natural products. Besides synthetically made products are relatively purer, as impurities that dilute the quality, are eliminated in the synthesis process.

However, from odor or essence point of view natural products are better and logically there is more demand of natural products than those derived synthetically.

In pure form essential oils are rarely used. These oils are marketed as a blend of the active ingredients (essential oils) and pure and dry ethanol. Essential oils are also marketed as bazaar oils after being dissolved in fixed oils as a mixture of essential oils and high boiling fixed oils. These are more commonly marketed in the Arabic bazars.

China and India used to occupy prominent places in the aromatic world during last 2000 – 3000 years. Of late China has marched forward enjoying a good global market. After introduction of synthetic flavor and fragrances in the form of essential oils India lost some ground. However, attempts are being made for popularization of Indian fine chemicals including aromatic oils. With R&D inputs in the field of aromatic or essential oils it is expected that India will regain its global market share in this field.

Important products isolated from plant sources are :

- Essential oils,

• Concretes, obtained by extraction with volatile solvents such as petroleum ether, benzene derivatives, or alcohols,
• Resinoids, produced by extraction of aromatic resins, masses, balsams and roots with volatile solvents,
• Absolutes, derived by treating concretes and resinoids with ethanol.

Starting materials for some of these products are shown vide Fig.1.1 [1,2].

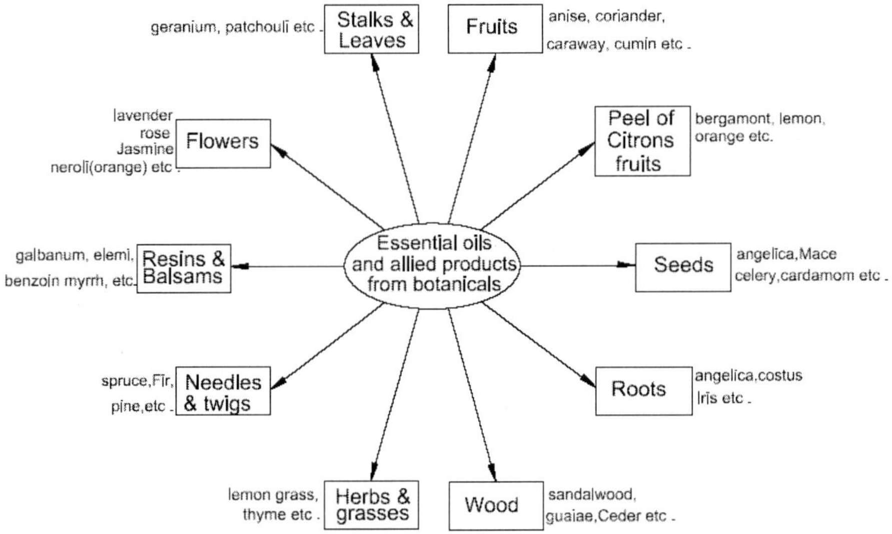

Fig. 1.1 Botanicas as sources of essential oils and allied products

1.3 Animal Origin – Sources of Fragrances

So far discussions were based on natural products based on plant origin that produces flavor and fragrances. Raw materials of animal origin, though rare and expensive, are still used in some cases. Following are some of the important sources,

• Ambergris, a metabolic product excreted by sperm whales,
• Musk, a glandular secretion of a hornless deer of Central Asia,
• Civet, a glandular secretion of the Civet cat,
• Castoreum, a glandular secretion of the beaver. [1]

1.4 Constituents of Essential Oils

Fig.1.2 shows various chemical compounds present in different essential oils [3].

Essential oils are probably connected with metabolism, fertilization and possibly protection from enemies. Any or all parts of the plant may contain oil. Buds, flowers, leaves, bark, stems fruits, seed, wood and in oleoresins exudates, are particularly rich in essential oils. These have been outlined earlier vide Fig.1.1.

Important aspects of essential oils, both natural and synthetic, will be discussed and examined, in the succeeding chapters.

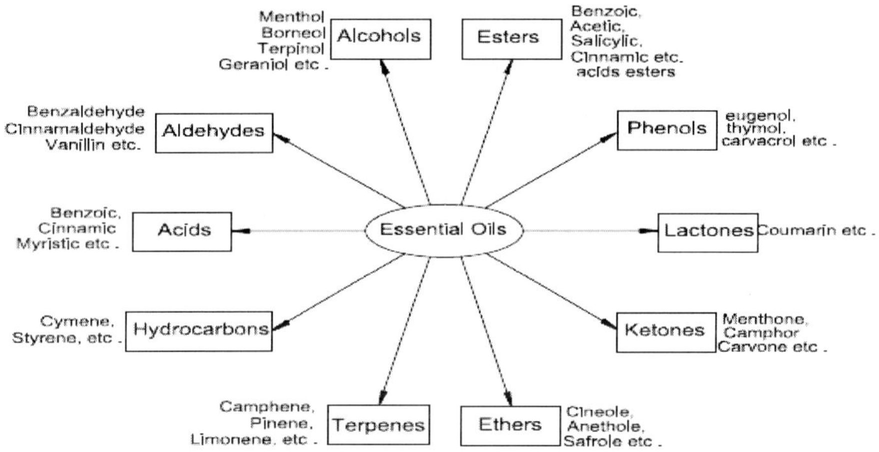

Fig. 1.2 Chemical compounds present in various essential oils

2

Isolation of Essential Oils

2.1 General

Essential oils based on natural extracts have been isolated as relatively pure products using traditional processes such as steam distillation, Enfluerage, maceration, expression and solvent extraction. However, steam distillation and solvent extraction processes are commercially attractive and other methods of isolation not being attractive any longer are being phased out.

The following chart Fig.2.1 summarizes the different ways in which aromatic oils are prepared [4].

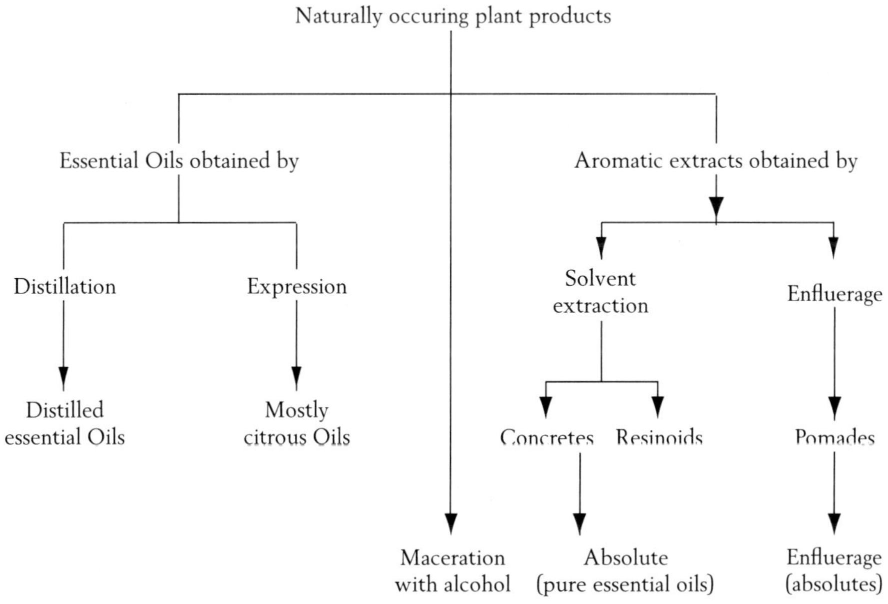

Fig. 2.1 Extraction of Essential Oils

2.2 Isolation of Natural Flavors and Fragrances

Numerous fragrances (and also flavors) are present in flowers, roots, stems, barks, leave and woods, Gums, balsams and oleo-resins are also valuable raw materials for extraction and isolation of active ingredients. Flowers owe their natural fragrances to the presence of traces of essential oils in their petals. Some of the flavor and fragrance containing natural products like Sandal wood, Cinnamon bark, Bay leaves similarly emit sweet, and in some cases pungent, odors even in natural form before extraction of important characteristic essential oils.

Processes for separation of essential oils from the odoriferous materials/botanicals mentioned earlier are briefly discussed.

Distillation and Extraction processes being techno-commercially most viable will be discussed in some details. To start with less important but conventional or traditional processes are briefly discussed here.

2.2.1 Maceration

The oil cells of fragrant flowers are ruptured by immersion in a hot fat or oil at about 60-70°C. This in turn absorbs the volatile or essential oils. Fat is then separated from spent flowers and reused for absorbing fragrant materials from the next batch or fresh flowers. The fat retained by flowers is recovered by hydraulic pressing. The resultant perfumed pomade is either marketed as such or is extracted with pure alcohol to yield extracts. Substantial loss of yield of desirable essential oils is reported.

2.2.2 Expression

For many citrous fruits such as lemon and orange peels this was an attractive traditional process in the past. For oils particularly the thermosensitive ones this was earlier used. More popular methods of isolation are the following:

- In the Ecuelle process, whole citrous fruits especially lemons, are rolled in hollow vessels covered inside with spikes to puncture oil cells which oozes out the essential oil into a collecting vessel for filtering to a clear oil.

- In the sponge process, the fruit is cut across the shorter axis, and the peel separated, steeped in water followed by hand pressing between sponges.

2.2.3 Enfluerage

Extraction of aromatic oils or perfumery chemicals by absorption of fragrances from flowers in contact with cold fats is the process of enfluerage. Fats should be saturated and odorless to avoid external contamination or adulteration of the desirable essential oils. Defluerage is caused by removing the spent flowers and fresh charges are made.

2.2.4 Distillation

Distillation is a unique physical process of separation which is caused because of differences in boiling points of one or more ingredients present in the mother stock.

The preparation of the botanical is extremely important and the method has to be adapted to the desired product.

If the material is to be water-distilled, it may be mixed with water in a holding tank before processing. However, specific treatment is required for each botanical and oil.

Steam or hydro distillation is the preferred method for producing essential oils employing either water, wet steam or dry steam.

In water distillation, the ground botanical is immersed in water which is heated to boiling. The oil and water vapors are passed into a condenser. The oil is separated from the water phase because of differences in densities of water and essential oil. Oil being lighter than water floats on the top and is easily separated. Oil can be dried and stored as a finished product. Water distillation is a mild but slow process, which yields a superior product from cloves or nutmegs, for instance.

The most common method of separation by distillation is the wet steam process. The ground botanical is placed on a grid in the still. There may be water below the grid or it may originate from the wet botanical. Steam is introduced from an outside source into the still. External heat may be applied to limit the built up of water. Steam distillation is a reasonably rapid method and is commonly used for most oils except those containing high concentration of low volatiles.

In case of dry steam distillation, the ground botanical is placed on a grid in the still. Direct steam is applied through a steam jacket avoiding water condensation. Ginger, Celery and Sandalwood, for instance are distilled by this method. Extraction of the distillate with a non polar solvent separates the organic phase from water. Evaporation of the low boiling solvent recovers the oil [5].

2.2.5 Extraction

Solvent extraction methods have been primarily employed for isolation of essential oils from botanicals which are heat sensitive such as Jasmine or tube rose or those containing an essential non-volatile constituent, e.g., piperine in black pepper. A proper liquid solvent should be low boiling, free of odor and impurities, and does not react with the extract. For food products, the solvent must be approved by the FDA, low boiling alkanes such as normal pentane or normal hexane for flower oils, Benzene or toluene for aromatic components, alcohol or acetone for phenolic components are selected preferred solvents.

The extraction is effected at room temperatures using solvents (b.p. 60-80°C) mentioned above. After extraction process is complete solvent is recovered between 88-93%. The concentrate is cooled, filtered and further concentrated and purified under vacuum.

2.2.6 Super Critical Fluid Extraction (SCFE)

Dense inert gases particularly CO_2 in the Super Critical phase has been effectively used for extraction of essential oils from the botanical. Fig.2.2 shows the pressure -temperature diagram of carbon dioxide [6].

SCFE using CO_2 beyond the critical point and high pressure is the most efficient and an acknowledged modern technology is gradually becoming very popular especially for a multi-product extraction system.

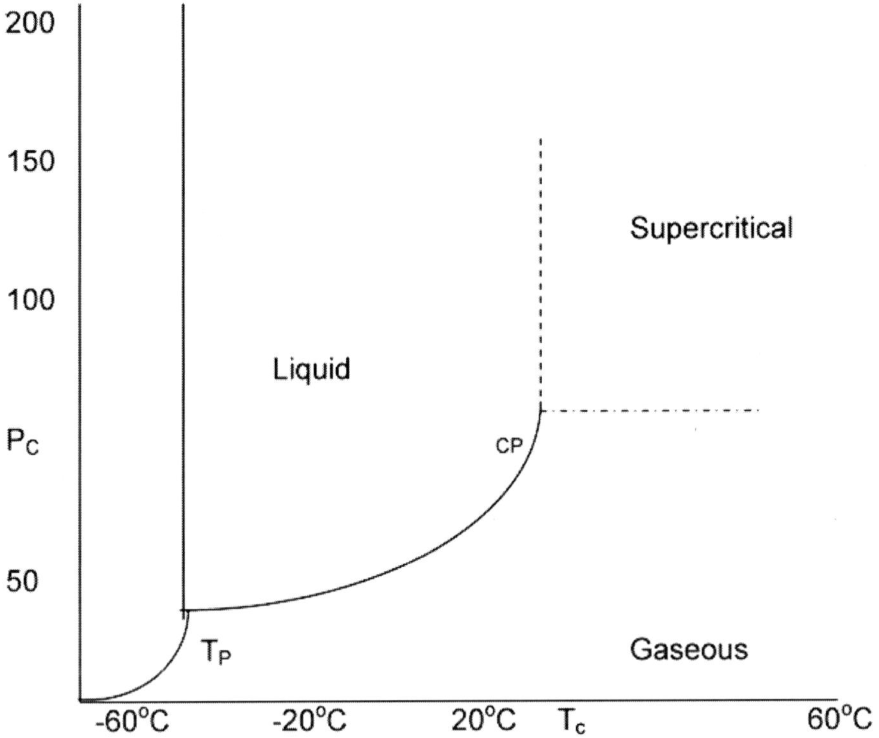

Fig. 2.2 Pressure/Temperature Diagram of Carbon Dioxide

T_p: Triple point -56.6°C, 5.2 bar; P_c and T_c: Critical Pressure and temperature; C_p: Critical point 31°C (temp.), 72.9 bar (pressure).

SCFE is shortly a two-step process that uses dense CO_2 in the gaseous condition above its critical temperature (31°C) and critical pressure above (73-74 bar).

Fig.2.3 briefly shows the process. The feed, generally ground solid, is charged into the extraction column through a high pressure pump (100-350 bar). After extraction at reduced temperature and pressure conditions, the extract, i.e., the product precipitates out in the separator. The extract free CO_2 stream is recycled to the extractor. SCFE process has wide application areas in the fields of spice oils and oleoresins, herbal medicines, flavors and fragrances encompassing important aromatic volatile oils etc. [7]

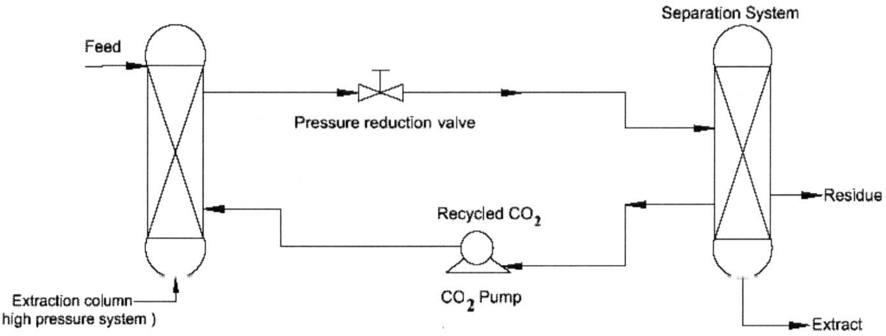

Fig. 2.3 Extraction Column (High Pressure System)
SCFE set up using high pressure CO_2 [Pressure 150-350 bar, Temp. 50-80°C, Time 3-5 hrs]

It is logical to presume that SCFE will not be limited to natural products extraction only. Gradually, the process will be extended to synthetic flavor and fragrance chemicals, food colors, pharmaceutical intermediates, where organic solvents are currently used for extraction purposes. Super critical CO_2 is a dense fluid and inert. The SCFE process will be suitable for extraction of synthetic aroma chemicals from a mixture of products and associated impure products. Yield of final product is higher, quality much better vis-à-vis conventional extraction processes. Use of SCFE for industrial/fine chemicals separation and extraction will ensure a clean process which is an eco-friendly system, and is a perfect example of green chemistry.

SCFE process has wide application in the field of some industrial and fine chemicals including spice oils and oleoresins, herbal medicines, flavors and fragrances, food colors and preservatives etc. Extraction of essential oils and allied products using SCFE process is now well-known. This is true not only in the extraction and recovery of essential oils not only from natural products but also for these essential oils or ingredients made by organic chemical synthesis.

This has became popular during last two decades or so. Extensive work is being carried out in various academic and industrial R&D establishments in USA, Canada, UK, Germany, Japan, Australia and other countries.

India is occupying a prestigious of place in development of SCFE technology and its commercialization. Work is being carried out in various academic and industrial R&D establishments particularly with a view to reducing capital cost for SCFE facilities. One may cite the commendable work done by the Chemical Engineering Department of the Indian Institute of Technology, Bombay, India, where a state-of-the-art multipurpose SCFE semi-commercial / Pilot plant has been developed, designed and made operational during the last 10-15 years. It has been claimed that based on the Institute's pioneering work, SCFE plants now cost 20-25% less than original conventional plants. The process has been successfully commercialized. One or two commercial plants have been already operating successfully. [7]

3

Some selected natural Essential Oils

3.1 Backdrop

Some 3000 oils have been identified from a large number of plant species and several hundreds have been commercialized. Many essential oils are well known and some are not as well known or might not have found commercial uses. Some important essential oils from natural botanical sources are shown vide Fig.3.1.

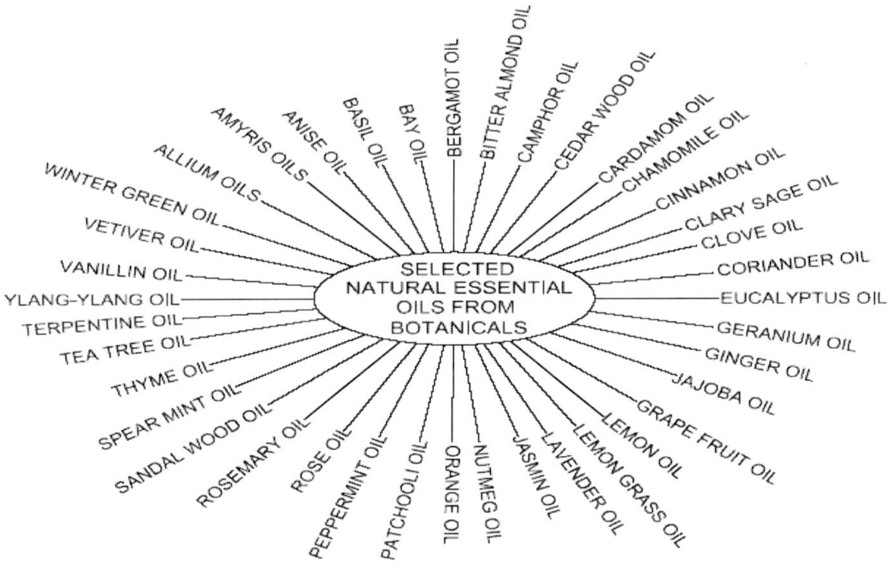

Fig. 3.1 Selected Essential Oils from Botanicals

Some of the important properties of selected essential oils from botanicals are listed vide Table 3.1. Their availability, uses etc. are also briefly elaborated.

3.2 Selected essential oils, their

Table 3.1 Selected Essential

Sr. No.	Product	CAS No.	Properties
1.	Allium oil	-	Obtained from garlic and onion Garlic oil Strong, pungent, garlic odor d_{25}^{25} 1.040-1,090 Diallyl disulfide {2179-57-9] is an essential odor component Onion oil Strongly pungent lasting onion odor d_{25}^{25} 1.050-1.135 Aliphatic Sulfur Compounds are responsible for onion odor
2.	Amyris oils	[8015-65-4]	Pale yellow to amber, slightly viscous liquid with a mild wood odor. d_{20}^{20} 0.946-0.978
3.	Anise oil	-	Colorless to pale yellow liquid with a sweet odor
4.	Anise, Star	-	Colorless to pale yellow liquid d_{4}^{15} 0.980-0.990
5.	Basil oil	[8015-73-4]	Light yellow liquid with a fresh, green, spicy odor
6.	Bay oil	-	Dark brown liquid with a strong spicy, clove like odor
7.	Bergamot oil	-	Green to greenish yellow liquid. The oil has a pleasant, fresh, sweet, fruity odor d_{20}^{20} 0.876-0.880
8.	Bitter Almond oil	-	Colorless to slightly yellow liquid with an intense, almond-like, cherry aroma

availability, properties, and uses

Oils and their properties [5, 10]

Availability	Uses
India, China, Russia, USA, Indonesia etc.	Mostly as flavor components, cooking etc. Seasoning mixture for the food industry
It is available as a substitute in West Indies for Sandalwood oil. Also produced in Caribbean area and around the Gulf of Mexico	Used in perfume compositions, mainly as a fixative
Arise plant is cultivated in Spain, Turkey and Egypt	For flavoring foods, beverages and oral hygiene products
Plant grows in South Western China and Vietnam	Baked goods, Candy, tobacco tooth paste, also in alcoholic beverage industry
Basil is cultivated in France, Mediterranean countries, Reunion, Comoros, Seychelles Islands.	Catsups, mustards, vinegar, and with cheeses and tomato-based recipes. Also in perfumery.
Bay trees are cultivated in northern South America and in the West Indies. Main cultivation area is the island of Dominica	Hair tonics, after shave lotion and also as a flavoring agent
Cultivated extensively in the Italian province of calabria, where soil and climatic conditions are favorable for Bergamot cultivation	Used in many perfumes and is an important component of lan-de cologne
The true bitter almond is cultivated in Morocco, Algeria, Spain, France	Almost exclusively in natural aroma compositions – for flavoring baked goods and soft drinks

Sr. No.	Product	CAS No.	Properties
9.	Camphor oils	[8008-51-3]	Both White Camphor oil and Brown camphor oil is available, colorless or almost colorless or pale yellow to brown liquid
10.	Cedar Wood oil	[68990-83-0]	Colorless to yellow liquid with a powerful herbaceous odor, resembling Thujone [546-80-5]
11.	Cardamom oil	[8000-66-6]	Colorless or very pale yellow liquid with an aromatic, penetrating, pungent odor
12.	Chamomile oil	-	Deep blue or bluish -green liquid. • Light yellow liquid. Sweet, woody balsamic odor
13.	Cinnmon oil	[8007-80-5]	Reddish-brown liquid with a sweet spicy, cinnamon-like odor
14.	Clary sage oil	[8016-63-5]	Pale yellow to yellow liquid with a herbaceous odor and a wine like bouquet
15.	Clove oil	[8000-34-8]	Yellow to brown, viscous liquid
16.	Coriander oil	[8008-52-4]	Almost colorless to Pale, yellow liquid, with a characteristic odor reminiscent of linalool. d^{20}_{20} 0.862-0.878
17.	Eucalyptus oil	-	Almost colorless, pale yellow, or greenish-yellow liquid. Australian oils are rich in Cineole content (80-85%), also some oils have fresh, lemon like odor containing citral

Availability	Uses
The oil is produced mostly in Taiwan and Southern China	Oil is used mainly in soaps and technical preparations. Marketed as Eucalyptus oil by the Chinese producers
Cedar wood oil is produced in Texas, USA, Mexico, China	Used mainly for perfuming soaps and other products, to convey woody notes to fragrances and also as fixatives
The plant is grown in India, Central America	Used primarily for seasoning foods and occasionally in perfumery
Available in three types • Blue chamomile oil. Plant is grown in central and Eastern Europe and Egypt,Argentina • English Chamomile oil Belgium, England, France and Hungary • Moroccan type	Primarily for aroma therapy, used as a component of perfumery blend.
The oil is produced in Sri Lanka, Seychelles, India, Madagascar and Comoro Islands	The oil is used as a fragrance in soaps. As a flavoring agent
The plant is cultivated in France, Hungary, Russia, USA. Spanish sage oils available in Spain, Almari, and Jaen.	Used in perfumes, toilet water, cosmetics and high grade soaps.
The trees are grown in Zambia, Tanzania, Madagascar, Indonesia and also in Sri Lanka and India.	Used in pharmaceuticals, baked goods, candies, mouth wash. As a flavor component, chemical synthesis
Coriander is mostly cultivated in Russia, Poland and Hungary, India etc.	Extensively used in Condiments, bakery goods, in seasoning mixtures, alcoholic beverages and also in perfume compositions.
These oils are produced commercially in Spain, Portugal, the Congo, South America, also in India, China, Australia	An ingredient of cough drops, mouth wash, dental preparations, inhalants, room sprays and medicated soaps. Also in perfumery.

Sr. No.	Product	CAS No.	Properties
18.	Geranium oil	-	The oil is an amber to greenish-yellow liquid with a characteristic rose like odor d^{20}_{20} 0.883-0.900 alcohol, citronellol / rhodinol 35-58%
19.	Ginger oil	[8007-08-7]	Light yellow to yellow liquid with a strong ginger-like odor but not so pungent
20.	Grape fruit oil	[8016-20-4]	A greenish-yellow liquid, with an odor resembling that of sweet orange oil but more herbaceous and bitter
21.	Jajoba oil	-	It has anti-inflammatory properties
22.	Lemon oil	-	Pale yellow to pale greenish-yellow liquid with a characteristic lemon peel odor. Composition depends on the variety of lemon and the country of origin
23.	Lemon grass oil	-	Indian and West Indian types differ in density d^{25}_{25} 0.869-0.894. Pale yellow to orange yellow liquid with a lemon-like odor resembling citral.
24.	Jasmine oil (absolute)	[8024-43-9]	The concrete is a brown to dark brown waxy mass with characteristic Jasmine odor. The absolute is generally a reddish brown liquid with a delicate jasmine odor.
25.	Lavender oil	-	Dark green color with a long lasting odor with a hay like spicy note
26.	Nutmeg oil	-	A colorless to pale yellow liquid with a pleasant spicy odor. d^{20}_{20} 0.883 - 0.927

Availability	Uses
The main cultivation areas of geranium are Reunion Madagascar, (Bourbon type) Morocco and Algeria	Used in soaps, perfumes especially for rose – like notes. Also chemical synthesis
Main ginger producers are India, Malaysia, Taiwan, Fiji	Used in beverages, baked goals, s a flavor component, ginger soaps
Grape fruit is cultivated in Israel, Florida, Texas and California in U.S.A.	As a food flavor in soft drinks, soft drink powders gelatin desserts, and candies. Also small amounts in lotions and colognes.
Available as liquid jajoba wax, USA, Europe, Indonesia, China, India and some African countries, produce this oil	It conditions and restores skin and hair and is used in shampoos and hair conditioners. A marvelous emollient.
The oil is produced in California, Argentina, Sicily, Cyprus, Brazil, Israel, Mexico, India, Spain, Australia etc.	An important food flavor in bakery goods, soft drinks, as a medicinal flavor and in USP properties
The oil is produced mainly in India. Also a different type is available in West Indies.	In addition to being processed into citral, it is used to certain extent for perfuming soap and house hold products.
Jasmine oil is produced in Morocco, France, Italy, Egypt, India, China, Taiwan and Japan. The concrete is made at the area of cultivation	The absolute is one of the most valuable blossom fragrances used in fine fragrances.
The plant grows wild but is cultivated in Southern France, Tanzania, Crimea	A stimulant and an ingredient of some USP pretreatment used extensively in cosmetics, lotions, soaps and shaving cream. Also an insect repellent.
The tree grows in Indonesia and West Indies	Used mainly in food flavorings and to a lesser extent in perfumery

Sr. No.	Product	CAS No.	Properties
27.	Mint oils (Peppermint oil Japanese)	[68917 -18 - 0]	Almost colorless to pale greenish-yellow liquid with a characteristic peppermint odor
28.	Patchouli oil	[8014-09-3]	Reddish brown to greenish brown, viscous liquid with a characteristic, woody balsamic odor
29.	Peppermint oils (Mint oils)	[8006-90-4]	Colorless to pale greenish-yellow liquid with a characteristic peppermint odor
30.	Orange oils	i) Orange oil, bitter [68916-04-1] ii) Orange oil, sweet - Nil	A pale yellow to yellowish brown liquid with mandarin-like odor with a bitter odor
31.	Rose oil	-	Yellow to greenish-yellow liquid with characteristic odor of rose blossoms and tea and honey notes
32.	Rosemary oil	[8000-25-7]	Colorless to Pale yellow liquid with a characteristic refreshing pleasant odor. d^{25}_{25} 0.894 – 0.912
33.	Sandalwood oil	[8006-87-9]	Slightly viscous, almost colorless to a yellow liquid with a characteristic sweet, woody long lasting odor. d^{20}_{20} 0.968 – 0.983
34.	Spearmint oils	[8008-79-5]	Colorless to yellow-green liquid with a fresh, caraway-minty odor. d^{20}_{20} 0.920 – 0.937
35.	Thyme oil	-	Red or reddish liquid with a strong characteristic, spicy-phenolic odor and a sharp lasting taste. d^{25}_{25} 0.910-0.935

Availability	Uses
The plant is cultivated in U.S.A., Brazil, Taiwan, China, Japan	Primarily used for isolation of menthol, in mouth washes, inhalants, tooth pastes, tobacco etc.
The patchouli shrub is cultivated primarily in Indonesia. The Philippines, Malay	Soaps and Perfumes
Peppermint is cultivated in USA, Russia, and Bulgaria. Production of European oil has decreased significantly.	Used primarily for flavoring tooth-paste, oral hygiene products, chewing gums. Smaller quantities are used for as a flavoring agent.
Produced mainly in a Mediterranean countries, Italy, Brazil and Spain	The oil is used as a flavor in bakery goods, soft drinks, candies and also as a component of perfumes.
Most of the commercial oils is produced in USA, West Indies, Spain, South Africa, Israel, India	A component of pharmaceutical, medicinal flavoring and recommended for treatment of bronchitis. Some USP preparations, as a flavor component
Most of the oil is produced in Bulgaria, Egypt and Turkey, Morocco	Primarily used as a high grade fragrance in cosmetics, toilet waters and perfumes. Also in flavoring of tobacco, and of food.
The main producers of rosemary oil are in Tunisia and Spain	Used widely in perfumery and in large amounts for perfuming bath foams, shampoos and hair
The trees grow in Mysore and Coorg (India). Also in Australia and West Indies	Used in soaps, cosmetics, perfumes and insense sticks.
The plant is cultivated mostly in Washington and Oregon (USA) and also in Germany, England, China.	Used as a food flavor in candies and chewing gums. Flavoring of tooth paste and in pharma items.
Thyme shrubs grow in France, Spain, Algeria, Morocco. Spain is the main producer of the oil.	Used mainly for flavoring foods and oral hygiene products. Also used in perfumery to create spicy, leathery notes. Has germicidal and antiseptic properties

Sr. No.	Product	CAS No.	Properties
36.	Tea tree oil	-	Pale to light yellow liquid with an earthy spicy odor d_{25}^{25} 0.900
37.	Turpentine oils	[8006-64-2]	Colorless liquids with a mild characteristic odor. Also have a phenolic odor d_{4}^{20} 0.855 - 0.870
38.	Vanilla (extracts)	-	Vanilla and phenol derivatives are primarily responsible for its aroma
39.	Vetiver oils	[8016-96-4]	Brown to reddish- brown, viscous liquid d_{20}^{20} 0.986-0.998
40.	Wintergreen oil	[60917-75-9]	Characteristic odor
41.	Ylang - Ylang-oil	[8006-81-3]	Height to dark yellow liquid with a floral, woody odor. d_{20}^{20} 0.910 to 0.956

3.2.1 Some additional herbs/shrubs or small trees as sources of popular essential oils. [4]

3.2.1.1 Celery

Celery is a garden plant with crisp juicy stalks, used predominantly as a spice for domestic uses and also in salads or as a vegetable. A familiar biennial plant, 30-60 cms high. Essential oils are extracted by stream distillation from the whole or crushed seeds.

Native to Southern Europe, extensively cultivated as a domestic vegetable. The oil is produced principally in India, and also the Netherlands, China Hungary and the USA.

Used for medicinal disorders a anti-oxidative anti-rheumatic and antiseptic.

Availability	Uses
Trees grow in Australia a Myrtacean	Used in perfumery for creating certain nuances and for earthy notes
USA and Canada, other major producers are Russia, Scandinavia and China also Portugal	Starting materials for flavor and fragrance components. Also as a raw material for chemicals. An insecticide ingredient, a paint solvent.
The main producers of vanilla pods are Madagascar, the Comoro islands, Reunion (Bourbon) Tahiti	Vanilla extracts are used extensively in chocolate, baked products and in ice-cream
Grows wild in Haiti, Cultivated in Reunion, Java and Brazil	Used in fine fragrances and long-lasting precious-wood notes. Also for chemical derivatives.
Eastern USA, particularly Pennsylvania and North Carolina	Used in liniment, mouthwash and toothpastes.
The oil is produced in Madagascar and the Comoro Islands.	Mostly in fine fragrances and also in soap perfumes.

3.2.1.2 Dill

Used since the earliest times as a medicinal and culinary herb. In Germany and Scandinavia especially, it is used with fish and cucumber, and the seeds baked in bread. Usually an annual or biennial herb with a smooth stem, feathery leaves and yellowish flowers.

Native to the Mediterranean and Black Sea regions, now cultivated in Europe, USA, China and India.

Used in aroma therapy/home use such as colic, dyspepsia, flatulence, indigestion having other widespread pharmaceutical uses. Also actions include anti-spasmodic, bactericidal, carminative, digestive uses.

3.2.1.3 Fennel

Biennial or perennial herb up to 2 m high, with feathery leaves and golden yellow flowers. Essential oil is extracted by steam distillation i) sweet

fennel oil from crushed seeds, and ii) bitter fennel oil from crushed seeds or the whole herb.

Bitter fennel is native to the Mediterranean region. It is cultivated extensively world wide. The main oil producers being Hungary, Bulgaria, Germany, France, Italy and India. Sweet fennel is grown principally in France, Italy and Greece.

Actions include Aperitif, anti-inflammatory, antimicrobial, antiseptic, anti-spasmodic etc. Used as culinary herbs or eaten as a vegetable.

3.2.1.4 Parslay

A biennial or short lived perennial herb up to 75 cm high with crinkly bright green foliage, small greenish-yellow flowers and producing small brown seeds.

Native to the Mediterranean region, especially Greece. It is cultivated extensively, mainly in California, Germany, France, Belgium, Hungary and parts of Asia.

Essential oil is extracted by steam distillation from i) the seed and ii) the herb. Also occasionally extracted from the roots.

Actions include anti-microbial, anti-rheumatic, antiseptic, astringent, carminative, diurectic etc.

- used extensively for seasoning and garnishing food.
- Also used as culinary herbs or eaten as a whole vegetable. Also used as a salad component in vegetables, fruits or also in fish etc.

3.3 Ayurveda, Herbal Medicines and Aroma Therapy [9,11]

The literal meaning of Ayurveda is the knowledge of like and longevity. Ayurveda encompasses the philosophy of purposeful and healthy living and a practical science. The highly integrated and intelligent mean with profound wisdom and erudition and compassion towards the predicament of the suffering humanity outlined remedies for various mental and physical illness.

Ayurveda is considered ads an upaveda (branch or section) of Rigveda, one of the four Vedas of ancient Hindu Civilization and Culture that

include Rigveda, Samaveda, Yajurveda and Atharvana veda. Atharva veda is the main source book of Ayurveda, and ancient systems of medicine and treatment.

Ayurveda is a rich science and proves effective for various ailments, such as chromic diseases like rheumatoid arthritis, osteo-arthritis, liver complaints. Ayurveda medicines and methodology, promise cure for various illness. It is in the fitness of things that the modes of administration of drugs practiced by Ayurveda nearly 5000 years ago have been recognized and certified as effective methods of drug administration by modern science.

Today there is greater understanding, awareness and interest in herbal medicines and aromatherapy, all over the world. Medical herbalism is a science and art of preparing natural herbal products and administering suitable drug preparations for specific symptoms and disorders. Medicinal and essential oils derived from specified herbs have been recommended as ayurvedic medicines to cure various ailments and disorders.

Aroma therapy based on aromatic plants and, more particularly aromatic essential oils, is being widely practiced. Some aspects of aroma therapy are discussed here in some details.

3.4 Aroma Therapy

Aroma therapy is an art and science of using botanicals or pant parts, mostly in the form of essential oils for well being of the humanity for removing or curing of diseases, physical, emotional, mental or spiritual imbalances. These essential or aromatic volatile oils cause remedial effects on various ailments. Needless to emphasize that aroma therapy had its origin in the teachings of the sages of ancient Hindu literature.

Our rishis (Indian Hindu Sages or Saints) were embodiment of plain living and high thinking. Numerous medicines or medicinal applications were discovered for the cure and welfare of the human beings. Aromatic plants, in particular aromatic essential oils, producing pleasing fragrances were practised and applied for cure and maintenance of good healths. Most of the diseases could be cured and taken care of

through applications of aroma therapy. Herbal medicines derived from not only aromatic volatile oils but also from herbs, spices, wood, roots etc. were made use of. Use of fragrances in the form of incense, perfumes, garlands, food condiments, cosmetic, massage oils, etc. to correct a great variety of physical and mental imbalances and to improve the general outlook of life etc. have been widely acclaimed.

Treatment of disorders commonly experienced through use of essential oils is shown vide Table 3.2.

Researchers in Europe and USA have been studying the precious essential oils and their blends for many years. Long before the term aroma therapy became popular, doctors in France were using essential oils to treat those wounded in wars when medical supplies ran short. There are several aroma therapy journals available that outline case histories and research.

However, as mentioned earlier aroma therapy was made popular in ancient India and to some extent in China.

3.5 Carrier Oils

More often than not pure essential oils are not used in undiluted form. Essential oils are too powerful to be used undiluted for use in bath, body, facial and massage purpose. Carrier oils or base oils are used to dilute pure essential oils suitable for aroma therapy. Some important base oils are discussed here in brief.

3.5.1 Almond oil

A fine, light, soothing and healing oil. Excellent for general use such as removal of make up and has very gold penetrating properties particularly good for dry sensitive skins.

3.5.2 Aloe Vera oil

It helps to maintain the skin's moisture balance and stimulates blood circulation. It tightens and rejuvenates the skin. Particularly good as a base for facial oils that treat dry, tired and infected skin.

3.5.3 Apricot Kernel oils

Full of vitamins and minerals this oily base is good for skin-care products for all types of skins. It is especially useful on sensitive and aging skin.

3.5.4 Avocado oil

Beneficial to all skins and contains Vitamins A and B and fatty acids.

3.5.5 Carrot Seed oil

Rich in Vitamins A, B_1, B_2 and C, minerals and also beta-carotene. This oil reduces skin sensitivity to sunlight and acts as a protection from sunlight.

3.5.6 Grapes seed oil

A light, easily absorbed, inexpensive oil; good for full body massage. Suits all skin types can be used undiluted.

3.5.7 Hazelnut oil

This oil is slightly astringent, so its application suits to oily damaged or combination skins. Hazelnut oil is also absorbed more easily than most other oils.

3.5.8 Jajoba oil

This liquid wax solidifies when allowed to cool. It is an excellent base for personal essences because it doesn't go off or become rancid as quickly as some of the others. Jajoba is nourishing to the skin and hair.

3.5.9 Pear oil

Is useful for dry skin. It contains Vitamin A and because of its viscosity aids absorption into the skin. It has a good penetration and therapeutic properties.

3.5.10 Wheat germ oil

Natural preservative. Very high in Vitamin E and also contains Vitamin A and D. It supports the skins natural process of regeneration, good for

dry and rough skin, makes skin very smooth. To dilute 25% in combination with another base oil and use.

3.6 Essential Oils to be Avoided for Aroma Therapy [11]

There are a number of essential oils that are best to avoid using altogether. These oils can be toxic and have little value for aroma therapy. Following are some of these oils:

Almond	Onion	Southern wood
Bitter Birch	Oregano	Tansy
Calamus	Pennyroyal	Wintergreen
Cassia	Rue	Wormseed
Lavender cotton	Sage	Wormwood
Mustard	Savory	

These oils have other important uses but because of toxicity are best avoided for skin treatment and direct uses for aroma therapy.

Table 3.2 Essential oils used for Aroma Therapy [9]

Sr. No.	Disorders/diseases	Essential oils for Treatment
1.	Anemia	Lavender, chamomile (Bath/massage)
2.	Loss of appetite	Chamomile (Bath/massage)
3.	Arthrists	Chamomile, cypress, Juniper, lavender, lemon, rosemary (Bath/massage)
4.	Bad breath	Lavender, tea tee (mouthwash)
5.	Baldness	Clary sage, lavender, rosemary, Ylang-ylang (use on scalp)
6.	Blood pressure	Clary sage, lavender, lemon, rosemary, Ylang-ylang (Bath/massage)
7.	Body odor	Clary sage, cypress, juniper, lemon (Bath/massage)
8.	Burns	Lavender, Chamomile, Eucalyptus, tea-tree (Bath/compress)

Sr. No.	Disorders/diseases	Essential oils for Treatment
9.	Poor Circulation	Cypress, juniper, lemon (Bath/massage)
10.	Colds	Eucalyptus, lavender, lemon (Bath, massage, rub on throat)
11.	Constipation	Rosemary, mint, orange, tangerine, Ylang-Ylang (bath, massage, rub on abdomen)
12.	Coughs	Cedarwood, eucalyptus, lavender, lemon, Sandalwood (Bath, inhalation, massage rub on throat and chest)
13.	Cuts	Eucalyptus, lavender, tea-tree (Compression etc.)
14.	Dandruff	Cedar wood, rosemary, lavender and tangerine (To use as hair oil, hair sine)
15.	Earache	Chamomile, lavender (compress etc.)
16.	Lack of energy	Orange (bath, massage etc.)
17.	Fatigue	Geranium, mint etc. (Bath, massage, room fragrance diffuser)
18.	Hair care	Chamomile, rosemary, geranium, lavender, sandal-wood (To use as hair oil, hair rinse)
19.	Hair loss	Clary sage, lavender, rosemary, Flange-Ylang (as hair oil)
20.	Headache	Clary Sage, lavender, rosemary, mint (Inhalation, massage)
21.	Indigestion	Chamomile, juniper, lemon, mint, orange, Sandal wood (Bath/massage etc.)
22.	Infections	Tea tree (as room fragrance diffuser)
23.	Insect bites	Chamomile, lavender, lemon tea tree etc.(Bath/com-press)
24.	Liver problems	Chamomile, geranium, rosemary, juniper, tangerine (Bath/massage, etc.)
25.	Respiratory problems	Cedar wood, eucalyptus mint, rosemary, sandal wood etc. (Bath, massage, compress, to rub chest and throat)
26.	Skin problems (dry, greasy or oily)	Cedar wood, lypress, Lemon mint, geranium, juniper, rosemary, Flang-Ylang etc. (Bath, compress, massage, etc.)
27.	Sprains	Chamomile (To use as cold compress)
28.	Sun burn	Lavender (Bath, massage, cold compress)

Sr. No.	Disorders/diseases	Essential oils for Treatment
29.	Female problems (menstrual)	Chamomile, clary sage, geranium, lavender, etc. (Bath/compress, massage, etc.)
30.	Toothache	Chamomile, mint, clove etc. Compress/mouthwash (Bath/massage or rub abdomens)
31.	Urination (Frequent)	Cypress, Bath, massage or oral abdomen

4

Downstream chemicals from Essential Oils - Natural and Synthetic

4.1 Backdrop

Essential/aromatic or volatile oils are naturally derived from botanicals and are a store house of various chemicals, fine chemicals, which are low volume and high value products. These natural chemicals have also been produced synthetically through systematic organic chemical synthesis. What nature, or more correctly Mother Nature, produced chemicals or active ingredients of flavor and fragrance chemicals, man has been successful in synthesizing these organic chemicals in various R&D organizations or laboratories in Research and educational institutions. Some of these organic chemicals, mostly fine chemicals, derived from natural essential oils and or synthetically produced are discussed here. [12, 13,14]

4.2 Chemicals from Essential Oils (Natural) and Synthetic Substitutes [10, 12, 14, 15]

There are hundreds of important downstream chemicals which are produced from natural essential oils. Many of them have since been synthesized and commercially produced in bulk quantities. Only some of these chemicals are discussed here by way of illustrations.

4.2.1 Anethole

Chief constituent of anise, star anise and fennel oils, responsible for characteristic odor and flavor of aniseed, produced mainly in Western Asia, Egypt. Also cultivated in Southern Europe and India, USA. 15-35% volatile anise oils are recovered from Anise seed. Anithole is also known as anise camphor, p-methoxy propenyl benzene or p-propenyl anisole.

CAS No.	[4180-23-8]
unspecified	[104-46-1]
Chemical formula	$CH_3CHCH\ C_6H_4OCH_3$

Chemical Structure

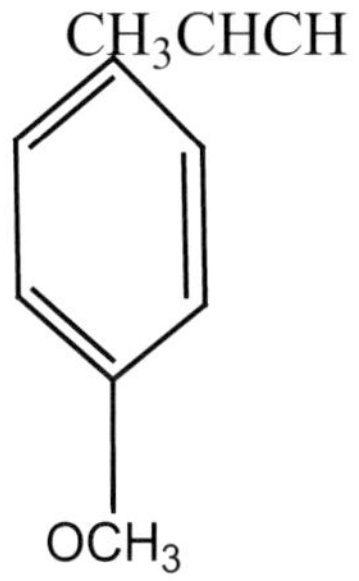

d_4^{20} 0.983-0.987

η_D^{20} 1.557-1.561

Anithole is obtained as white crystals, having sweet taste. Main source: via crystallization of anise or fennel oils, also synthetically from p-cresol.

4.2.2 Anisaldehyde

CAS No.	[123-11-5]
Chemical formula	$C_8H_8O_2$

Chemical Structure

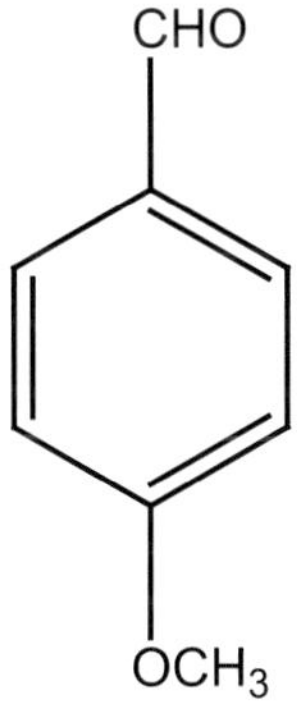

M.W. 136

M.P. 0° C

B.P. 248° C

d_4^{20} 1.119-1.122

η_D^{20} 1.570-1.572

Anisldehyde, popularly known as para anisic aldehyde or p-anisaldehyde or p-methoxy-benzaldehyde or even as anbepine, an important

perfumery material was extracted from natural essential or volatile oils of fennel oils or even as a minor ingredient of anise seed oil and also from Acacia farneisiana or cassia flower scented babul, sweet acacia etc.

Isolation of anisaldehyde from natural sources i.e., essential oils containing anisaldehyde has been commercially not viable. However, considering the multiplicity of uses of the product it was considered prudent to commercially produce the product synthetically from p-cresol. This has been discussed in details in an earlier publication. [7]

4.2.3 Borneol

CAS No.	[507-70-0]
Chemical formula	$C_{10}H_{18}O$

Chemical Structure

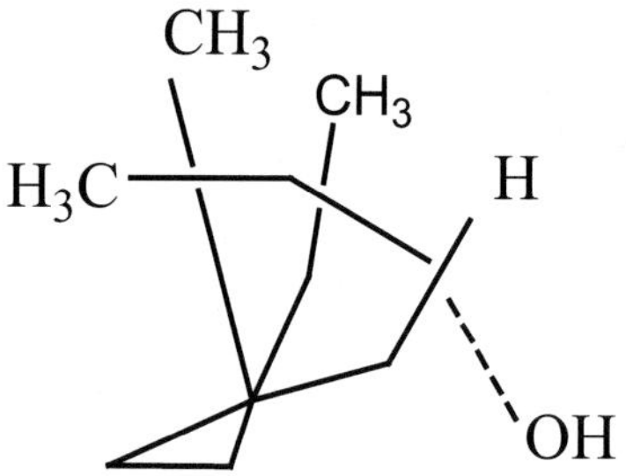

M.W. 154,25
M.P. 208° C
B.P. 212° C
d_4^{20} 1.011

- Hexagonal plates from pet. Ether
- Sublimes but is less volatile than camphor
- Insoluble in water, soluble in alcohol, petroleum ether, acetone etc.
- Has a peculiar peppery odor and burning taste, somewhat resembling odor of mint.
- Naturally occurs in a species of tree in Borneo and Sumatra (Indonesia). Also can be produced through organic chemical synthesis from camphor (by hydrogen reduction) or from α-pinene.
- Used for perfumery and flavors, incense and also in pharmaceuticals.

4.2.3.1 Bornyl acetate

Colorless liquid, pine needle odor, very slightly soluble in water, soluble in alcohol, ether, soluble in most fixed oils and in mineral oil, Preparation of the d-form by acylation of d-borneol. Uses: Perfumery, flavoring, nitrocellulose solvent.

CAS No.	[76-49-3]	
Chemical formulas	$C_{12}H_{20}O_2$	$CH_3COOC_{10}H_{17}$

M.W. 196.29

M.P. 29° C (crystals)

B.P. 220-224° C

d_4^{20} 0.978

4.2.4 Camphor

CAS No.	[76-22-2]
Chemical formula	$C_{10}H_{16}O$

Chemical Structure

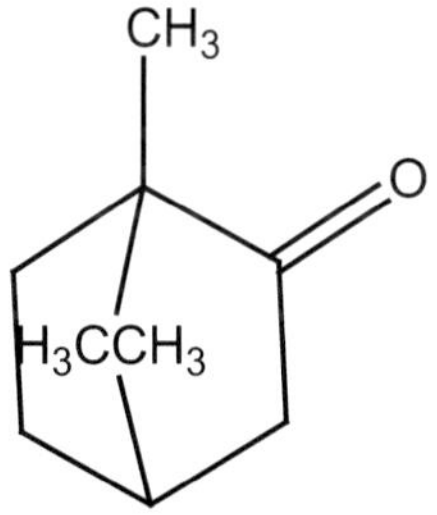

M.W. 152.23

d_4^{25} 0.992

A ketone occurring naturally in the wood of the camphor tree cinnamomum camphora naturally occurs in both d-form and ℓ-form. Natural camphor is produced via steam distillation of camphor-tree wood and crystallization.

Synthetic camphor, most of which is optically inactive, is made from pinene, which is converted to camphene, and thereafter by treatment with acetic acid and nitrobenzene is converted to camphor (see Fig.4.1).

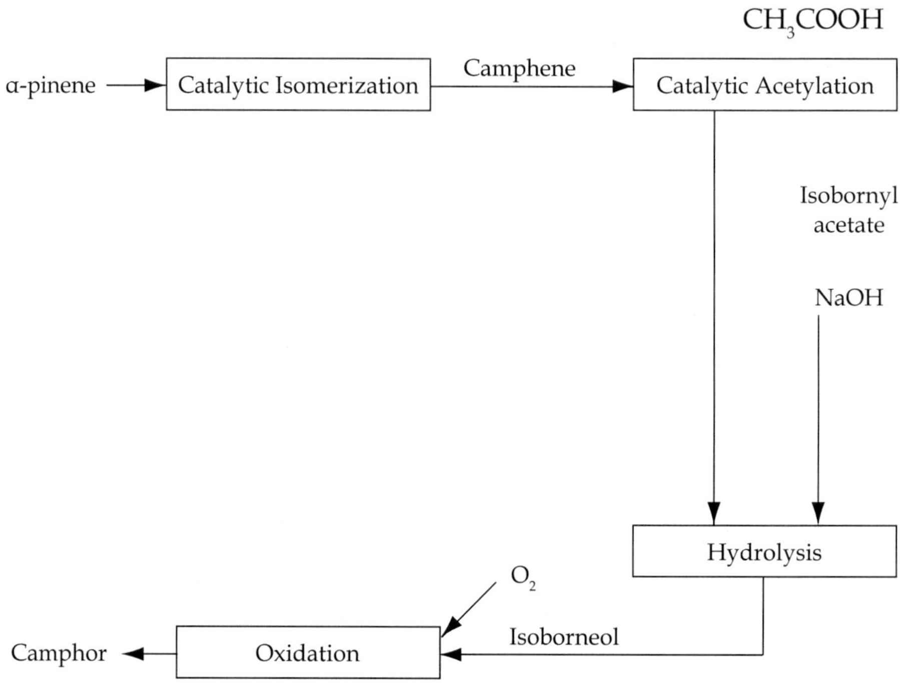

Fig. 4.1 Block diagram for Camphor from α-pinene

Camphor is obtained as white or colorless crystals, or as colorless to white translucent masses.

- Camphor has characteristic fragrant and penetrating odor with pungent, aromatic taste.
- Commercially two grades are available, Technical grade m.p. 163-168°C, usp m.p. 174-179°C.

Uses: used as an odorant and flavorant, also as a pharmaceutical product. Finds applications as an insecticide primarily as a moth repellent.

4.2.5 Chavicol

CAS No.	[501-92-8]	
Chemical formula	$C_9H_{10}O$	$CH_3H_5C_6H_4OH$

Chemical Structure

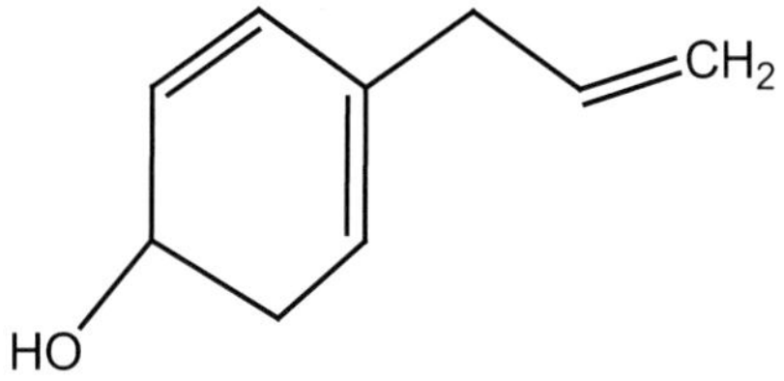

M.P. 16° C

B.P. 230° C

d_4^{18} 1.033

also known as 4-(2-propenyl) phenol or p-allylphenol

- Liquid
- Occurs in many essential oils, found together with terpenes in volatile betel oils

4.2.6 Carvone

CAS No.	[99-49-O]
Chemical formula	$C_{10}H_{14}O$

Chemical Structures

$CH_3\overline{C\ CH\ CH_2CH[C(CH_3)CH_2\ CH_2\ C}O$

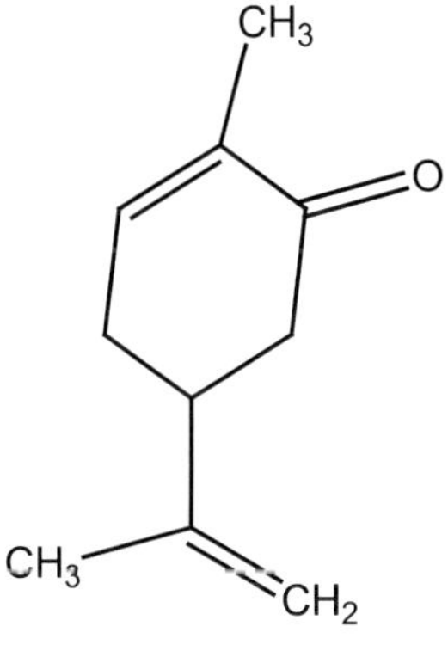

M.W. 150.22

B.P. 227-230° C

d_4^{20} 0.960

η_D^{18} 1.4999

- d-carvone is found in caraway seed oil, Dill seed oil,
- d-carvone is also isolated from mandarin peel oils, l-carrone is found in spear mint and Kuromoji oils. Also prepared from α-pinene, may be synthesized from d-limonene,
- soluble in alcohol, ether, chloroform, propylene glycol, insoluble in water.

Properties: Pale-yellowish or colorless liquid with a strong characteristic odor.

Uses: flavoring, liquors, perfumery, soaps.

4.2.7 Cinnamic aldehyde

CAS No.	[104-55-2]
Chemical formula	C_9H_8O

Chemical Structure

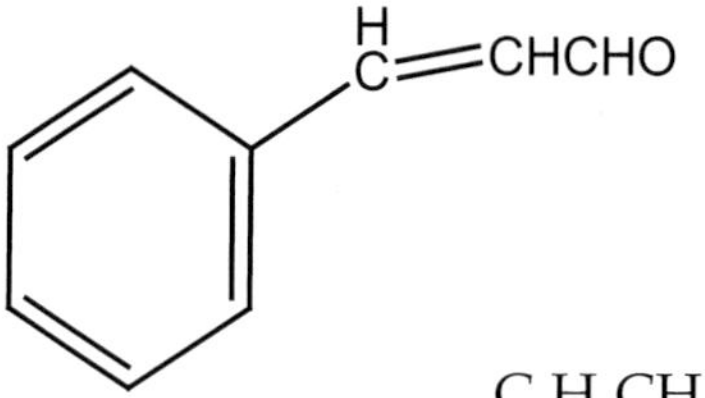

M.W. 132.16

d_4^{15} 1.048-1.052

η_D^{20} 1.618-1.623

$C_6H_5CH = CHCHO$

- Yellowish oil, odor of cinnamon, sweet taste
- Soluble in five volumes of 60% alcohol, very slightly soluble in water, combustible.

Derivatives: Produced from Ceylon and Chinese (Saigon) cinnamon oils. Also cinnamon is cultivated in tropical Africa, America and Asia synthetically produced by condensation of benzaldehyde and acetaldehyde.

Uses: Flavors and perfumery products, Cinnamaldehyde is used in many compositions for creating spicy and oriental notes (soap perfumes).

4.2.7.1 Cinnamyl acetate

It is a good fixative and is used in blossom compositions and for oriental notes. In aroma compositions, it is used for cinnamon-fruity effects.

CAS No.	[21040-45-9]
Chemical formula	$C_{11}H_{12}O_2$

Chemical Structure

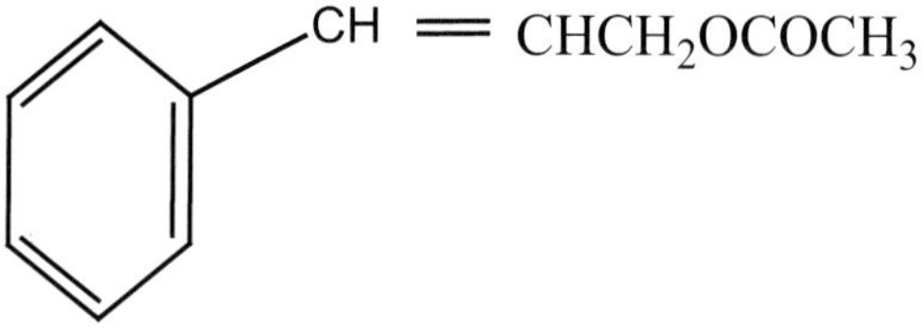

M.W. 176.21

d_4^{22} 1.0520

η_D^{20} 1.5420

- Is an ester of cinnamic alcohol of any importance.
- trans-cinnamyl acetate occurs in cassia oil and is a colorless liquid with a sweet-flowery-fruity, slightly balsamic odor.

4.2.7.2 Cinnamic alcohol

CAS No.	[104-51-1]
Chemical formula	$C_9H_{10}O$

Chemical Structure

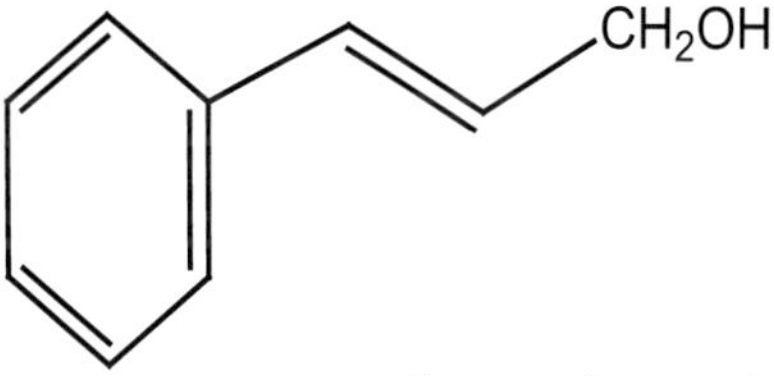

trans-cinnamic alcohol

M.W. 134.18

M.P. 34

d_4^{20} 1.0440

η_D^{20} 1.5819

Exists in both Cis and trans forms.

- Both isomers occur in nature, tarns-isomer is far more abundant and is present, for example, in styrax oil.
- Trans-cinnamic alcohol is a colorless, crystalline solid with a hyacinth-like balsamic odor.
- Synthetically produced by reduction of cinnamaldehyde.

Uses: Cinnamic alcohol is valuable in perfumery for its odor and fixative properties. It is a component of many flower compositions (lilac,

hyacinth, and lily of the valley). In aromas, the alcohol is used for cinnamon notes and for rounding off fruit aromas. Also used in synthesis of pharmaceutical products.

4.2.8 Citral

CAS No.	[5392–40-5]
Chemical formula	$C_{10}H_{16}O$

M.W. 152.24

Occurs as cis and trans isomers

Natural citral is nearly always a mixture of the two isomers. It occurs in lemongrass oil (upto 85%) and in Litsec Cubela oil (upto 75%), and in small amounts in many other essential oils.

- The citrals are colorless to slightly yellowish liquids, with an odor reminiscent of lemon.

Smaller quantities are isolated from essential oils. Bulk quantities required for synthesis of Vitamin A are synthetically produced.

- Cital is isolated from lemongrass oil and from Litsea Cubepa oil. Recovery is made by distillation.
- Citral is produced in bulk quantities from geraniol by synthesis.

Uses: Citral is very important for aroma compositions because of its strong lemon odor. In perfumery it is used in neutral media due to its discoloration, oxidation etc. It is used in synthesis of ionones as intermediates for Vitamin A synthesis.

4.2.9 Citronellal

Citronellal occurs in essential oils in its (+) and (-) forms, often together with the racemate. Pure citronellal is a colorless liquid with a refreshing odor, reminiscent of balm mint. Citronellal is still isolated from essential oils in considerable quantities. Also produced synthetically. Citronellal is obtained from Citronella oils by fractional distillation. (±) Citronellal is isolated from Eucalyptus Citriodara oils when necessary; it is purified by using an addition compound such as bisulfite derivative. Citronel-

CAS No.	[106-23-0]
Chemical formula	$C_{10}H_{18}O$

Chemical Structure

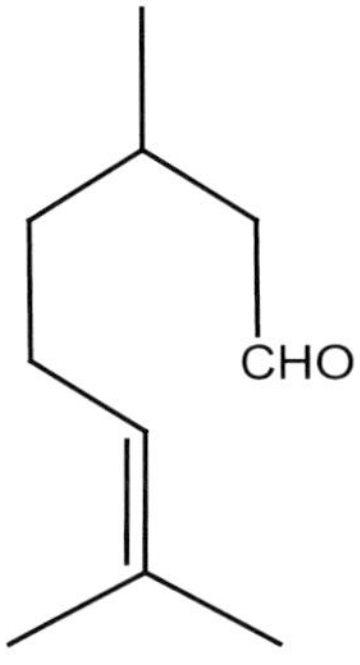

M.W. 154.25

d_4^{20} 0.851

η_D^{20} 1.4477

lal is also synthesized from Geranial or nerol, also via synthesis from citral. Uses: Citronellal is used in limited quantities for soaps and detergents. Main use is an organic chemical synthesis.

4.2.10 Eugenol

CAS No.	[97-53-0]
Chemical formula	$C_{10}H_{12}O_2$

Chemical Structure

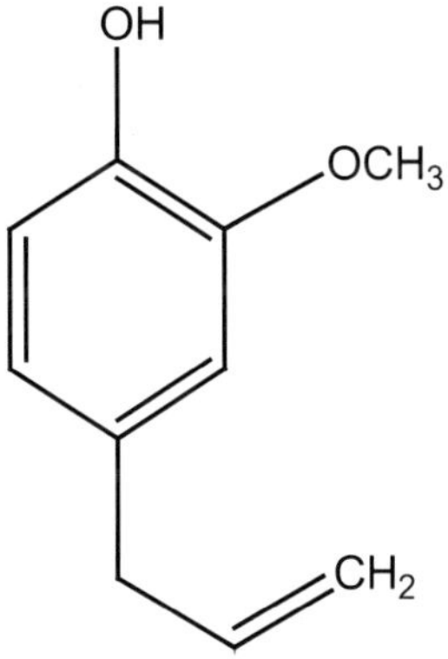

M.W. 164.20

d_4^{20} 1.0837

η_D^{20} 1.5726

- Obtained from many natural sources, colorless or pale yellow liquid.
- Odor of cloves, spicy, pungent taste clove leaf oil and cinnamon leaf oil may contain >90%

Uses: Perfumery

- oil of cloves
- Vanillin making
- Insect attractant
- In dentistry, it is used as an antiseptic.

Since sufficient eugenol can be isolated from cheap essential oils, synthesis is not industrially important. Eugenol is still preferentially isolated from clove leaf and cinnamon leaf oil by extraction with sodium hydroxide solution.

4.2.11 Eucalyptol

CAS No.	[470-82-6] also known as: Cineole or Cajepulol
Chemical formula	$C_{10}H_{18}O$

Chemical Structure

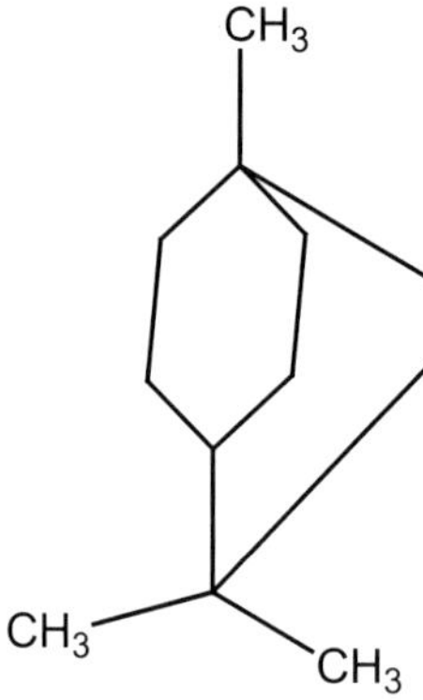

M.W. 154.25

d_4^{20} 0.9267

η_D^{20} 1.4586

Occurs in many terpene-containing essential oils. For example eucalyptus oil contains upto 85%, laurel leaf oil contains upto 70%. It is a colorless liquid with a characteristic odor. Slightly reminiscent of camphor.

Eucalyptol or Cineole is a fragrance material obtained exclusively by isolation from essential oils especially eucalyptus oils. Recovered by fractional distillation as in Spain and further purified by crystallization of eucalyptol/cineole rich eucalyptus oil fraction.

It is miscible with alcohol, chloroform, ether, glacial acetic acid etc. It is used in large quantities in fragrances and flavors, mostly as oral hygiene products.

4.2.12 Estragole

CAS No.	[140-67-0] also known as: chavicol methyl ether or estragol
Chemical formula	$C_{10}H_{12}O$

Chemical Structure

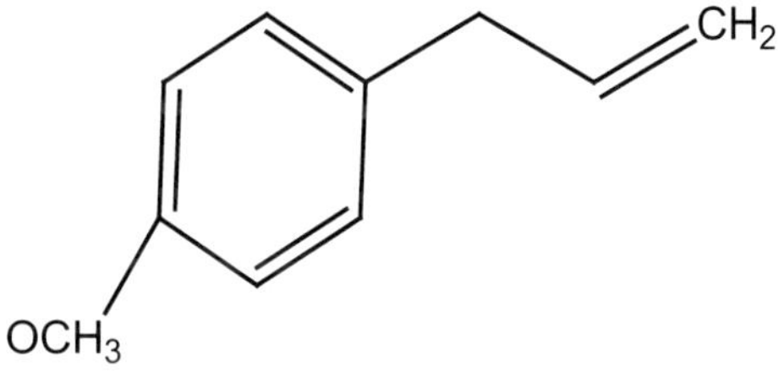

M.W. 148.20

d_4^{21} 0.9645

Liquid, soluble in alcohol, chloroform, forms azeotropic mixture with water. Main constituent of terragon oil is also known as estragon oil. It occurs also in pine oil and in American terpentine oil.

Estragole is used as a component of some perfumes and as a flavor component in foods and liquors.

4.2.13 Geraniol

Geraniol is isolated from Citronella oils and from palmarosa oils. Fractional distillation from Java Citronella oil yields a fraction containing approximately 60% geraniol, as well as citronellol.

Geraniol is frequently used fragrance material. In flavor compositions, geraniol is used to accentuate citrons notes.

CAS No.	[106-24-1] also known as: lemonol
Chemical formula	$C_{10}H_{18}O$

Chemical Structure

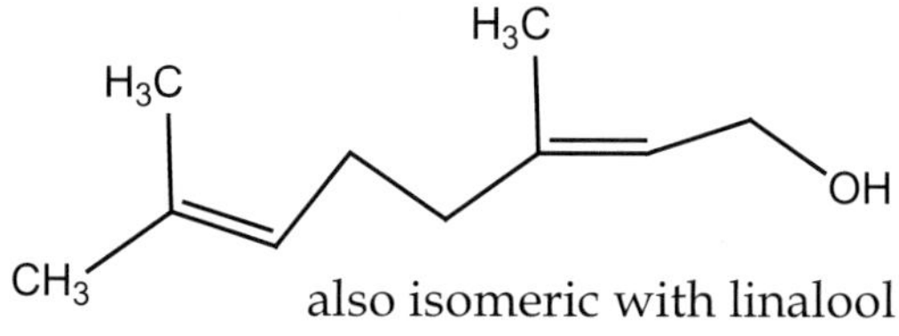

M.W. 154.25

d_4^{20} 0.8894

η_D^{20} 1.4777

also isomeric with linalool

- Occurs in all terpene-containing essential oils, frequently as an ester.
- Palmarose oil contains 70-85% geraniol, geranium and rose oils also contain large quantities.
- Geraniol is a colorless liquid with a flowery-roselike odor.

4.2.14 Coumarin

CAS No.	[91-64-5] also known as: 1,2-benzopyrone
Chemical formula	$C_9H_6O_2$

Chemical Structure

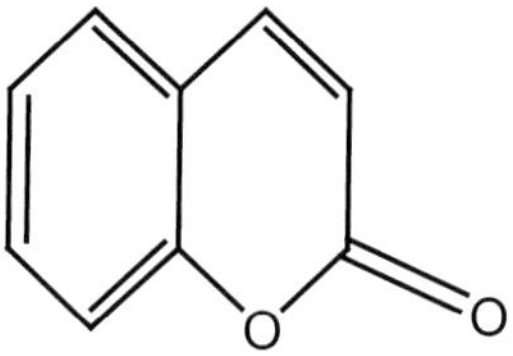

M.W. 146.15

M.P. 69.73° C

d_4^{20} 0.935

Crystalline Powder

Coumarin is one of the most important aroma chemicals having unique characteristics not only because of its hay-like bitter sweet odor but also because it is perfume fixative.

It is widely distributed in the plant kingdom though principally it is produced synthetically. Mostly the synthetic material is produced either from o-cresol, phenol and also salicylaldehyde.

Uses: Mainly used as a flavoring agent and as an intermediate in the Pharmaceutical industry. It has a pleasant fragrant odor resembling vanilla.

4.2.15 Thymol

CAS No.	[89-83-8] also known as: 2-isopropyl 5-methyl phenol
Chemical formula	$C_{10}H_{14}O$

Chemical Structure

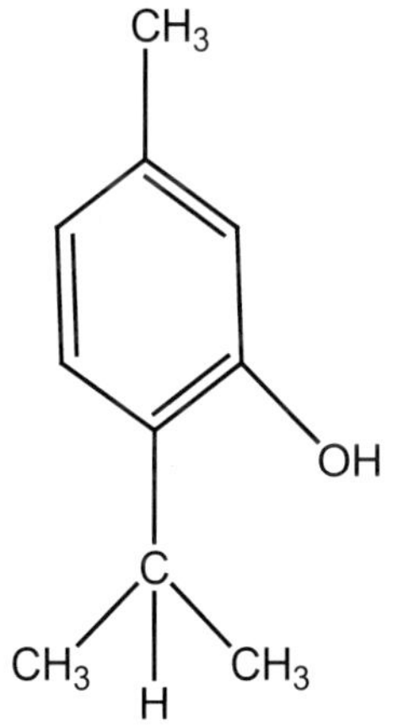

M.W.	150.22
M.P.	49-51° C
d_4^{20}	0.9756
η_D^{20}	1.5227

Colorless crystals, spicy herbal, slightly medicinal odor.

Thymol is present in volatile oil of thyme and ajowan, mint seeds from which it may be extracted with sodium hydroxide, liberated from the sodium compound with acid and purified by crystallization. It has a woody, burnt, smoky odor. Used as a flavoring agent but its most important use is for Menthol production.

4.2.16 Menthol

CAS No.	[89-78-1] also known as: p-methan-3-o ℓ -hexahydrothymol
Chemical formula	$C_{10}H_{20}O$

Chemical Structure

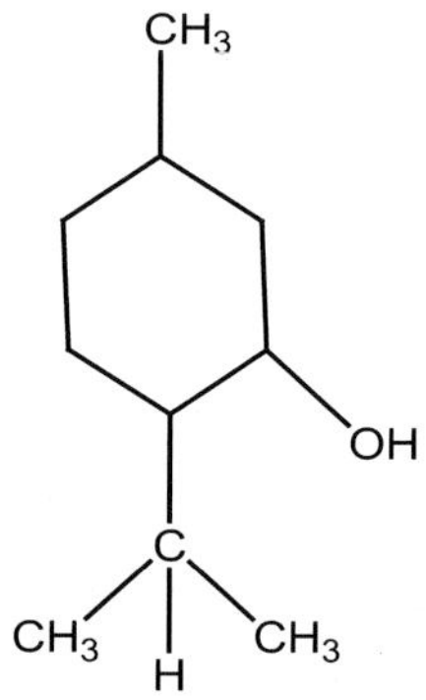

M.W. 156.27

M.P. 43° C/ (-form)

B.P. 216° C

d_4^{20} 0.890

η_D^{20} 1.5227

White crystals, cooling odor and taste; congealing temperature 27-28°C.

(-) Menthol is a white, waxy crystalline compound having a strong odor of peppermint. Its main source is oil of peppermint from the plant Mentha arvensis. It is an ingredient of decongestant ointments and nascal sprays and is also used in flavor toothpaste and cigarettes.

During recent years, natural menthol is supplemented with synthetic menthol. One popular route is synthesis from citronellal via isomerization and hydrogenation. Commercially, however, the more economic route is from m-cresol through thymol hydrogenation.

4.2.17 Vanillin

Vanillin is a natural product and can be found as a glucoside in vanilla beans at concentration of nearly 2%. It can be extracted with water, alcohol and other organic solvents. The best known source of vanillin is the vanilla plant, vanilla platifolia A, which belongs to the Orchid family. It is cultivated mainly in Mexico, Madagascor, Reunion, Java and Tahiti.

Vanillin is a universally popular flavoring agent. Vanillin exhibits white crystalline needless having sweetish smell.

CAS No.	[121-33-5] also known as: 3, methoxy-4-hydroxy benzaldehyde or Vanillie aldehyde
Chemical formula	$C_8H_8O_3$

Chemical Structure

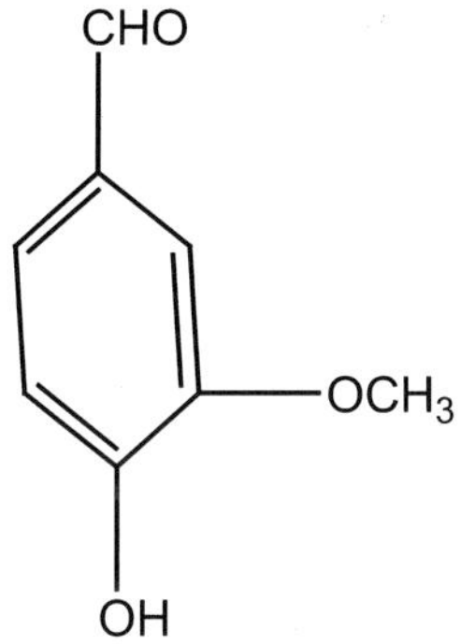

M.W. 152
M.P. 82° C/ (-form)
B.P. 285° C
d_4^{20} 1.056

Vanillin is produced synthetically from Eugenol, Guaicol and recently from p-cresol. Major uses of Vanillin:

- In flavor formulations, as a sweetener or as a flavor enhancer
- In butter, chocolate and all types of fruit flavors, root beer, cream soda etc.
- In baking
- In chemical synthesis.

4.2.18 Linalool

Linalool occurs as one of the main components in many essential oils. (-) Linalool [126-90-9] occurs at a concentration of 80-85% in skin oils from cinnamomum camphora, cayenne linaloe oil contains approximately, 80% + Linalool [126-91-0] makes up 60-70% of coriander oil.

CAS No.	[78-70-6]
Chemical formula	$C_{10}H_{18}O$ or $(CH_3)_2\ C = CH\text{-}CH_2\text{-}$ $CH_2C(CH_3)(OH)CH{=}CH_2$

Chemical Structure

M.W. 154,25

d_4^{20} 0.8700

η_D^{20} 1.4616

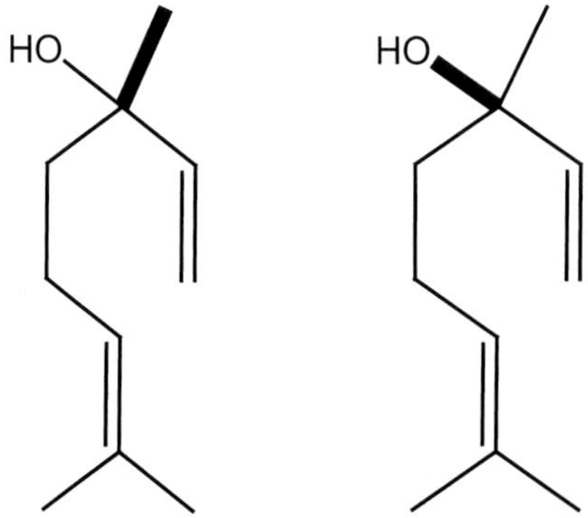

(+) Linalool (-) Linalool

Properties: A colourless liquid with a flowery-fresh odor, reminiscent of lily of the valley. Together with its esters linalool is one of the most frequently used fragrance substances and is produced in bulk quantities.

- Linlool is isolated from essential oils by fractional distillation. Some oils are rosewood oil, shin oil and coriander oil.
- Also synthesized from alpha-pinene/beta-pinene.

4.2.18.1 Linalyl acetate

CAS No.	[115-95-7]	
Chemical formula	$C_{12}H_{20}O_2$	$CH_3COOC_{10}H_{17}$

M.W. 196.29

d_4^{20} 0.895

- Most valuable constituent of bergamot and lavender oils.
- Also found in many other volatile oils

- Liquid bergamot odor
- Miscible with alcohol, ether
- Mostly used as a component of perfumes.

4.2.19 Raspberry ketone

CAS No.	[5471-51-2]
Chemical formula	$C_{10}H_{12}O_2$

Chemical Structure

M.W. 164.2
M.P. 82-83° C
B.P. 200° C at 760mm pressure

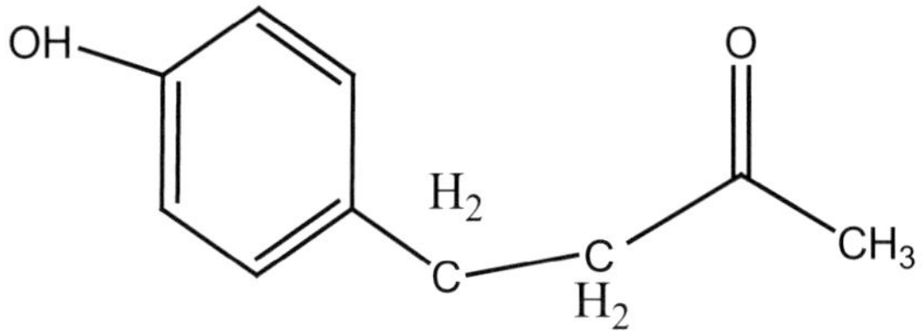

- Highly characteristic component of raspberry
- Forms colorless crystals (M.P. 82-83° C) with a sweet-fruity odor strongly reminiscent of raspberries.

While a large number of volatile compounds contribute to the flavor of raspberries, most are present at less than 10 ppm.

- Compounds include alcohols, acids, esters, ketones and other hydrocarbons. However, a particular ketone 1-(4-hydroxy-phenyl)-3-butanone has an odor very much like raspberry. This particular ketone, i.e., raspberry ketone, is sold under trade names Frambinon (Dragoco) and oxyphenylon (IFF).
- It may not be out of place to mention that ketones are partly or wholly responsible for the odor of many natural products such as camphor, raspberries, coffee, musk, caraway, butter, rue and jasmine.

Natural occurrence: Beef, pineapple, Raspberry, strawberry.

Perfumery uses: Babe, Blackberry, and Blassc', Bluebery.

Synthetic raspberry ketone is a high value low volume product. Very popular in Europe, particularly U.K. and Germany and also in North America.

- Traditionally raspberry ketone has been synthetically made from phenol and methyl-vinyl-ketone.
- Now made from p-hydroxy benzaldehyde from p-cresol oxidation.

For details one may refer to an earlier publication. [7]

4.2.20 Longifolene

CAS No.	[475-20-7]
Chemical formula	$C_{15}H_{24}$

Chemical Structure

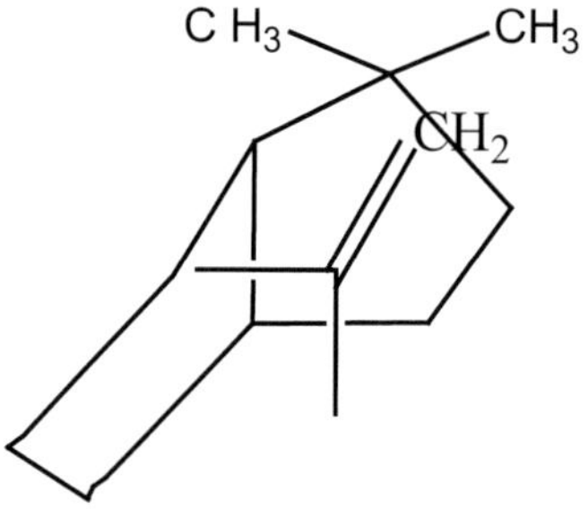

M.W. 204.35

d_4^{18} 0.9319

η_D^{20} 1.5040

- Viscous oil
- Soluble in benzene and other aromatic hydrocarbons
- Also in Indian terpentine oils which are commercially produced from the Himalayan pine.

Used in specific perfumes, pharmaceutical Formulations along with menthol and terpenes.

4.2.21 Alpha-Terpineol

Alpha-Terpineol is colorless, crystalline solid, smelling of lilac. The most important commercial grade of terpeneol consists of a liquid mixture of isomers, that contains mostly α-terpineol and a considerable amount

of γ-terpineol. α-terpineol occurs in many essential oils, however, only small quantities are isolated by fractional distillation of pine oils (long leaf pine oils).
Uses: Terpineol with its typical lilac odor is one of the most frequently used fragrance compounds. It is also used in soaps and cosmetics.

CAS No.	[98-55-5] also known as: 1-p-menthen-8-ol
Chemical formula	$C_{10}H_{18}O$

Chemical Structure

M.W.	154.25
M.P.	40-41° C
M.P.	35° C (racemate)
d_4^{20}	0.9338 (Liquid product)
d_4^{20}	0.9357
η_D^{20}	1.479

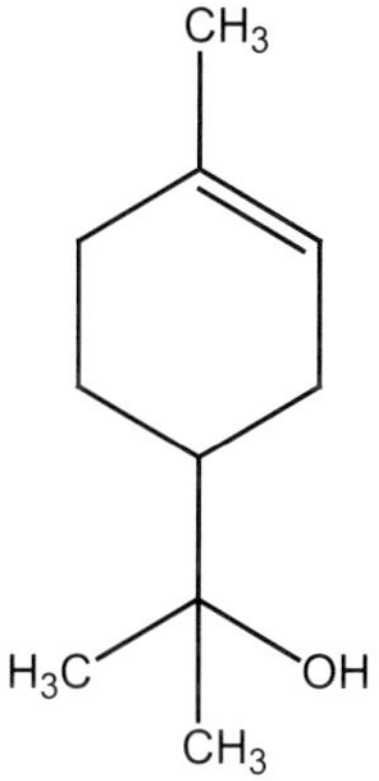

4.2.22 Rose Oxide

Rose Oxide occurs in small quantities, mainly the levorotatory cis form, in essential oils (such as Bulgarian rose oil and geranium oil). Commercially synthetic products are either optically active or inactive mixtures of the cis and trans isomers. They are colorless liquids with a strong odor reminiscent of geranium oil and carrot leaves. Rose oxide is usually prepared from citronellol, which can be converted into a mixture of two allyl hydroperoxides. (-) Citronellol as the starting material Yields approximately a 1:1 mixture of (-) cis and (-) trans – rose oxide.
Uses: Rose oxide is used in rose and geranium perfumes.

CAS No.	[16409-43-1] also known as: 4-methyl-2-(-2methyl-1-proponyl) tetrahydropyran
Chemical formula	$C_{10}H_{18}O$

Chemical Structure

M.W. 154.25

d_4^{20} 0.875

η_D^{20} 1.4570

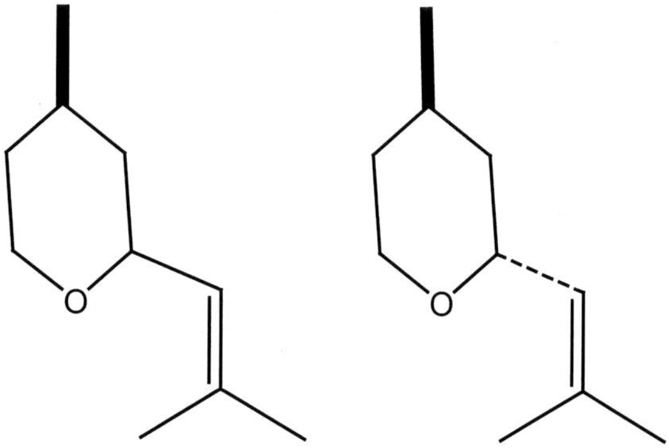

(-) Cis-rose oxide (-) trans rose oxide

4.2.23 Ionones [10]

The C_{13} ketones α- and β-ionone are cyclic terpenoids that occur in many essential oils. Also r-ionone has been also detected though this does not yet been isolated from botanicals.

Physical properties of the three isomers of ionones are shown in the following table:

Name	Formula	CAS no.	MW	d_4^{20}	η_D^{20}	Odor
α-Ionone	$C_{13}H_{20}O$	[127-41-3]	19230	0.9319	1.4982	Sweet-floral, reminiscent of violets
β-Ionone	$C_{13}H_{20}O$	[79-77-6]	192.30	0.9461	1.5202	Reminiscent of cedar wood, violet like
γ-Ionone	$C_{13}H_{20}O$	[79-76-5]	192.30	0.9317	1.4985	Violet-like with woody resinous tonality

Structural formulas are as follows:

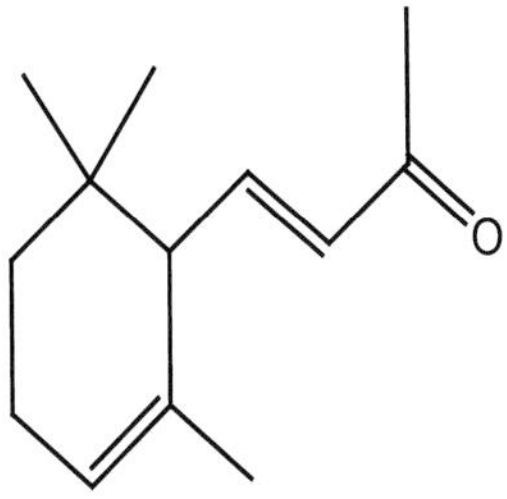

α-lohone

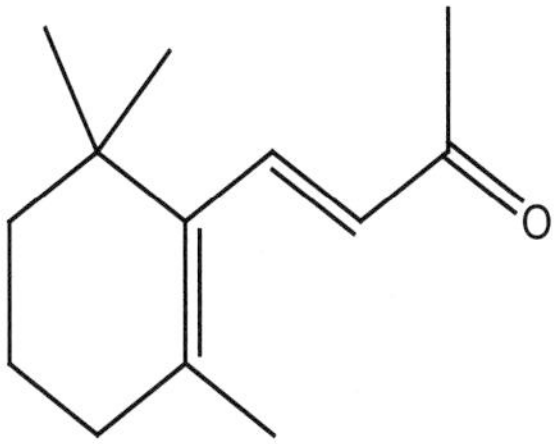

β-lonone

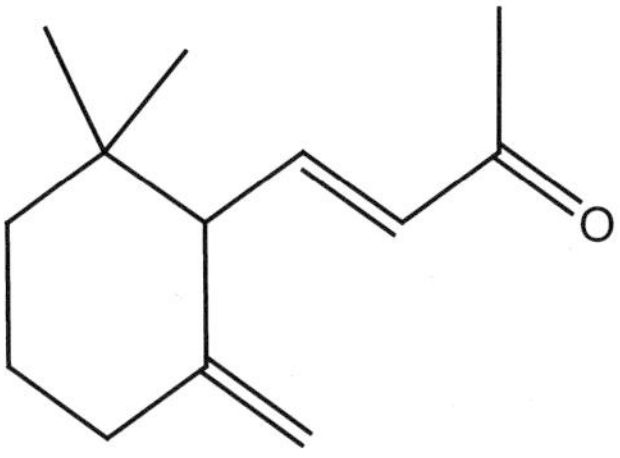

γ - lonone

β-Ionone is commercially most important and is discussed in some details. α-Ionone is a highly valued fragrance material.

4.2.23.1 Beta-Ionone [14]

A pale yellow liquid having a boiling point of 130°C at 10.0 mm of mercury has a characteristic violet-like odor. The refractive index at 20°C (η_D^{20}) is 1.5190 to 1.5215.

Raw materials:

1. Natural mode from Lemongrass oils, pine based terpenes are also used for natural product.
2. Synthetic beta-ionone starts from acetylene. Other chemicals are Acetone, Caustic soda and Sulfuric acid.

Uses: It is the base for manufacture of Vitamin A and is also used in perfumery. Its incorporation in fragrance formulations imparts a pleasing sweetness which adds depth and warmth to the formulation.

4.3 Aroma Chemicals from Essential Oils - Indian Scenario [16]

India occupies a pride of place in the world of essential oils and chemical derivatives.

The share of India in global essential oils market used to hover around 0.5-1.5% till seventies in the last century. But today Indian share is between 16-17%. This is mostly because of mint oil production.

Terpene chemicals like Citral, Citronellal, Geraniol, Hydroxy Centranellal, Ionone, Methyl Ionone, Linalool, Menthol etc. were consid-

ered as major aroma chemicals, the raw materials for the manufacture of fragrances and flavors. The source for the manufacture of these major terpene chemicals were essential oils like Lemongrass oil, Citronella oil, Rose oils, Mentha Arvenesis etc.

There are about 30 major aroma chemicals whose individual annual consumption in fragrances and flavor industry is more than 1000 MT. Out of these 30 major chemicals more than 50% are terpene chemicals.

The present status of Lemongrass oil and Sandal wood industry is not promising. On the other hand the development of agro technology of citronella oil, Mint oil, Basil oil, Geranium oil, Patchouli oil is a success story.

The development of agro technology for production of menthol and other terpene chemicals like Linalool, Linyl acetate and anethole ex Indian Basil oil has enabled Indian manufacturers to produce the base essential oils and derived aroma chemicals at competitive prices for supply to the global market. [18]

4.4 Essential Oils from the Forestry Sector [19]

India bounded on three sides by the Seas and mighty Himalayas in the north has forests which are sources of a number of botanicals that produce essential oils and downstream aroma chemicals.

Some of the important botanical sources for essential oils are as follows:

Leaf: Eucalyptus Globulus, Patchouli, Lemongrass, Basil, Cinnamon, Clove

Woods: Sandalwood, Agar wood, Cedar wood, Turpentine

Roots: Nalerian, Jatamansi, Vetiver, Nagarmotha, Angelica, Costus, Calamus

Flowers: Lavender, Clary sage, Mimosa, Bakul, Champa, St. John's Wort, Lotus, Keroa.

Bark: Cinnamon, pine

Seeds: Linaloe Berry

From economic point of view, the most important are a) Sandal wood, b) Agar wood, c) Turpentine, d) Eucalyptus, e) Cedar wood

Once these essential oils are identified and isolated, extracted and are obtained in pure form, downstreams aroma chemicals can easily be isolated and purified. A vision 2010 has been also presented in this paper. [19]

Results and extrapolations can be summarized as follows:

- Geraniol from palmarosa / Jamrosa oil
- Eugenol, Anethole, Linalool, Methyl charicol, Methyl exgenol, Methyl Cinnamate from Basil oil.
- Eugenol from clove leaf oil,
- Citronellal, Citronellol and Geranial from Citronella oil
- Citral, Vitamin A from Lemon grass oil
- Eucalpytol from Eucalyptus Globulus oil
- Citronellal, Citronelloe from Eucalyptus Citrodora oil
- L-Methone from Mentha piprita oil
- Thymol, Menthone from Mentha Awansis oil
- L-carvon from spear mint oil

It has been projected that by 2010. 20000 MT of essential oils will be produced from forest sources and along with downstream derivatives a turn- over of 350,000 US dollars will be achieved.

5

Essential Oils and Perfumes

5.1 Backdrop

Essential oils, whether they are ingredients of flavors or fragrances, are either aromatic compounds or their derivatives, and are the most important components of perfumes. These are fragrant essences used to scent the body, cosmetics, and candles. More than 100 natural aromatic chemicals may be blended from a range of 60000 flowers, leaves, fruits, seeds, woods, barks, resins and roots, combined with fixatives and synthetics. Favored ingredients include balsam, Civet (from the African Civet Cats), hyacinth, jasmine, lily of the valley musk (from the musk deer), orange blossom, rose, and tuberose. [17]

Use of perfumes and perfumery materials is perhaps as old as the human civilization. Use of perfumes in some form or other was reported not only in the celestial empire of gods but also in ancient civilizations of Egypt, Rome, India, Greeks, and China etc.

Egyptian beauties many thousand years ago were known to have made prolific uses of scents or perfumes and even some colors, in all parts of their bodies. Indians were using perfumes or perfumed materials in all festivals including religious ceremonies and for rites and rituals. Similarly Greeks and Chinese were using perfumes in various occasions. Soon at a later stage Romans also started using various perfumes.

5.2 Raw Materials

The raw materials used in perfumery are essential oils, floral oils, extracts from plants and also from animal secretions, fragrance substances isolated from natural products and also purely synthetic aroma chemicals. These were briefly discussed in the introductory chapter.

5.2.1 Fixatives

Fixatives are used to equalize the vapor pressures, and thus the volatilities of the raw materials in perfume oil as well as to increase the tenacity. Natural fixatives are resinoids (berzoin, labdanum, myrrh, olibanum, storax and tolu balsam) and animal products ambergris, castoreum, musk and civet). Similarly use of synthetic fixatives such as benzyl salicylate, ambroxide, and also virtually odorless solvents with very low vapor pressures such as benzyl benzoate, diethyl phthalate, triethyl citrate etc.

5.2.2 Solvents

Extremely pure ethanol has been predominantly used as a solvent. Sometimes Ethanol is diluted with distilled water to bring down the concentration.

Some weak odorous synthetic fixatives are used as solvents for alcohol-free perfumes especially bazaar oils which are perfumed oils sold in bazaars in Arabian countries as perfumed oils. These are currently also used by the Muslim populations of Pakistan and India, and other countries.

5.3 Essential Oils and Perfumery Blending [20]

Perfumery blending is a very unique art where in there are no fixed principles or restrictions while composing a perfumery blend. It may be a very complex formulation or even a good perfume may be created with simple and a few odoriferous materials consisting of aromatic chemicals (natural or synthetic) essential oils and resinoids, ultimate formulation should be pleasing, appealing and acceptable.

Essential oils in perfumery can be used in neat form at a pre-determined ratio or they may be processed and used for the blended and final perfume.

Three major variations for use of perfumery blends have been focused.

- Various essential oils may be used in neat form
- Oils should preferably be terpine less
- Oils are processed and isolates and derivatives are used.

With the advancement of Science and Technology the quality and availability of odoriferous materials also improved. Art and Science of blending were also perfected. Natural essential oils and to a lesser extent, synthetic flavors and fragrance materials, play very significant roles in the process of blending.

A perfumer has at his disposal two hundred odd essential oils. Some of these oils are used liberally, some sparingly, and some in small proportions to get exotic effects. Again differences are made for male perfumes and female perfumes.

5.4 Creation of Perfumes [21, 22, 23]

The creation of perfumes is a commercial art practiced in the medium of odorous substances. It is highly specialized and individualistic. A widely varying composition of ingredients, numbering in thousands and varying in chemical composition is available to perfumers.

Table 5.1 shows some commonly used odor descriptions grouped into five generally classifications.

Table 5.1
Perfumery descriptions [21]

Floral	Citrous	Woody	Green	Fruity
Carnation lilac	Bergamot	Cedar	Basil	Apple
Chrysdanthemum Lilly	Grape fruit	Fir	Cucumber	Apricot
Gardenia marigold	Lemon	Hickory	Grass	Banana
Honeysuckle mugnet	Lime	Patchouoli	Parsley	Black currant
Lyacinth narcissus	Mandarin	Pine	Rhubarb	Cherry
Iris orange flower	Orange Tangerine	Sandal	String Bean Violet	Fig grape melon
Jasmine rose	verbena		Violet watercress	Pineapple Prune, melon
Jonquil	Violet			Raspberry
Lavender	Mimosa			Strawberry

Odor perception and description are highly subjective in nature. There however has been generally agreed-upon odor vocabulary that is used to characterize individual ingredients and finished fragrances.

5.5 Fine Fragrances

Fine fragrances must work on the skin and blend with body odor. These should be pleasant, diffusive and long-lasting, and have the quality of genuine beauty and should distinguish them from each other. For most fine fragrances, the perfumes are themselves the products. They are sold to the consumer at various concentrations in alcoholic or aqueous-alcoholic solutions.

5.6 Perfume Compositions [1]

Perfume compositions prepared in the form of perfume oils are dissolved in ethanol, cooled to approximate 0°C, filtered packaged, stored, and marketed. Perfume may be, for example a 10-20% by volume, solution of perfume oil in approximately 90-95% vol ethanol. An eau de toilette or eau de- perfume contains approximately 7-10% perfume oil in 80-90 vol% ethanol, an eau de cologne upto 4 vol% perfume oil in 70-85 vol% ethanol.

A fragrance can be divided into three parts:

- Top note or initial fragrance impression
- Body note or middle note
- End note

There are many formulations available which are tailor made, or on many occasions finalized on the expert's advice.

A rose base composition for instance, is shown vide Table 5.2

The many perfumes now available on the market can be classified according to the main fragrance concepts such as

- Jasmine concept based on its constituents
- Lilac such as Terpineol and Anisaldehyde and
- rose such as phenethyl alcohol Geraniol, Citronellol, rose oil etc. and also lily of the valley (or its chemical constituents), iris violet or its

constituents such as Ionones, methylionones etc. and clove (that includes Eugenol, Iso-eugenol, clove oil etc.).

Table 5.2
Rose base

Composition	Quantity gms		Compound
Top note	15	→	2-undecylene aldehyde (10%)
	5	→	ω-decenol
	5	→	phenyl acetaldehyde dimethyl acetate
	20	→	Decanal 1% in diethyl phthalate
Body note	300	→	Phenethyl alcohol
	200	→	Citronellol
	100	→	Geraniol
	20	→	Gerenyl acetate
	160	→	Phenethyl diethyl acetate
	25	→	Nerol
	20	→	Farnesol
	25	→	Nerolidol
	5	→	Eugenol
Fixation	20	→	Trichloromethyl phenyl Carbonyl acetate
	30	→	Cinnamic alcohol
	50	→	Phenethyl phenylacetate
	1000 gms		

6

CONCLUSIONS

6.1 Synopsis

From the dawn of civilization many thousands of years ago man realized that 'man cannot live by bread alone'. Soon man understood the value of so many other things. For instance man understood the value of botanicals i.e., plant parts such as flowers, fruits, bark, roots etc. The concept of smell i.e., essence and its value in totality was understood and man started using plant parts for various end uses.

Flavor and fragrance or the active ingredients of botanicals was soon established and applied for diverse end uses. The word 'essential oil' was understood and conceptualized at a later date. There were wide spread applications in the ancient civilizations of Egypt, Greece, India, China and later on in the ancient Roman civilizations.

Once humanity understood the role of God / gods and religion, people realized that offerings and pujas were to be made in the name of gods in ancient days and later on to the modern days. People understood that offerings to gods ere to be made in pure minds and pure bodies. So it became common to make use of scents and perfumes in concentrated form including as incense sticks (or agarbattis) while worshipping gods. Even for religious rites and rituals and even during birth and death of individuals from times immemorials.

Use of various fragrances or essential oils by the Egyptian women many thousand years ago is well-known. Some ancient Egyptian beauties started making use of scents or fragrances and even flavors and colors to decorate their entire bodies to make themselves more appealing attractive and even seducing their heroes, kings and other personalities.

With the advancement of Science and Technology during last few hundred years people started understanding the chemistry of essential oils from botanicals and soon started isolation of active ingredients and the composition there of. Some 6000-7000 years ago perfumes and perfumery chemicals encompassing flavors and fragrances attracted attention in India and were used in Hindu temples and also for religious rites and also personal sophistication. The earliest references to such material were to be found in the Vedas. In particular, Sandal wood, Camphor, Saffron etc. were mentioned in connection with some rites.

Indeed in the fascinating world of essential oils, flavors and fragrances, perfumes and perfumery chemicals many things happened and are still happening. Some aspects have been discussed in some details in the preceding chapters.

Till about 150 years ago practically all the products used in perfume compositions were of natural origin and were isolated from botanicals. Mother Nature was lavish in creating hundreds of plants, herbs and shrubs in all parts of the world. Time passed. Through systematic scientific research men learned the art of synthesizing the active ingredients perfumes and perfumery products consisted of. Today hundreds of synthetic products analogous to the natural ones have been created. In some cases synthetically prepared blends of perfumes were even better and commercially more attractive than those procured from natural sources since impurities present in natural perfumes could be eliminated.

6.2 Current Scenario [7, 15]

Though there are hundreds of companies engaged in the business of aroma chemicals, it is strange that the top eight full service companies - manufacturers in all three categories of flavors, fragrances and aroma chemicals - enjoy two thirds of the world' total business. A few hundred small companies account for only one third of the business.

Table 6.1 shows the market shares of the top multinational companies in flavor, fragrance and other aroma chemicals. Total business in the year 2007 is estimated at US$16 billion.

Table 6.1

Market shares of flavor, fragrances and other aroma chemicals*

Sr. No.	Company	Market Share
1	IFF	14.2%
2	Givaudan-Roure	11.9%
3	Quest	9.4%
4	Haarman & Reimer	8.9%
5	Firmenich	7.2%
6	Takasago	6.2%
7	Bush Boake Allen	3.8%
8	Dragaco	3.1%
	Subtotal	64.7%
	All others	35.3%
	Total	100%

*Figures are as of 2000 – 2001.

Thanks to internationalization of business most of the key players in the business of flavors and fragrances know exactly where they should establish manufacturing facilities for which of the aroma chemicals. Needless to emphasize that China and India – the two largest countries in the world with more than 2 billion people will be two major business centers. Both these countries have abundance of natural resources in the field of aroma chemicals and both have been suppliers of chemical, pharmaceutical and dyestuff intermediates. New technologies are also being developed or implemented at regular intervals.

6.3 Indian Scenario

India holds a very important place among the oldest civilized countries. More than 7000 years ago India emerged as a civilized country. It is hardly a surprise than flavors and fragrances in general and perfumes in particular were used by the Indians long ago. The concepts of harbalism and aroma therapy were introduced by the great saints and sages

(Moonies) in ancient times. Medicinal effects of many essential oils were understood and applications were made for cure of many diseases.

India is a very rich source of botanicals or essential oils from plant sources. Jasmine, rose, Sandalwood etc. are grown abundantly in many parts of the country. India also produces many scientists of international repute every year. It is no wonder that India is very advanced in the field of organic chemical synthesis in fine chemicals and synthetic essential oils. There is little doubt that India has emerged as a leading supplier of flavor and fragrance chemicals as reported by the Flavors and Fragrances Association of India or FAFAI 2000.

India has one of the largest cresols complexes in the world. All the cresol isomers and its derivatives such as para anisic aldehyde have wide applications as components of perfumes.

- Perfumery grades of p-anisic aldehyde and para anisic alcohol made form p-cresol are used by many of the top aroma chemicals companies such as IFF, Givaudan, BASF, H&R, Dragaco, S.H. Kelkar etc.
- Some of the esters and ethers of para cresol and its derivatives have become very popular in the manufacture of synthetic perfumes. Prominent among them are para cresyl methyl ether, p-cresyl phenyl acetate etc. Some of them are very popular in the manufacture of Indian Agarbattis (incense sticks).
- Thymol and Menthol which were traditionally made from botanicals are now synthetically made from meta-cresol. BAYER A.G. of Germany is the largest manufacturer of synthetic thymol and menthol.
- Synthetic musk used as perfumery fixative musk ambrette is made from meta cresol.
- Coumarin, a major perfumery fixative is made from ortho-cresol.

Many of the synthetic flavors and fragrances chemicals have been discussed in some details in Chapter 4.

6.4 Future Outlook

Essential or volatile aromatic oils and perfumery compositions as their blends had a beginning thousands of years ago but the subject is so fas-

cinating that new perfumes are created almost every year in every country. A good perfume made from both natural sources or from synthetic blends is always popular. Synthetic products are cheaper that those made from botanicals. But perfumes made from natural sources or botanicals are more pleasant and naturally more attractive.

Big multinational companies having multiple locations in different countries have contributed in a big way for development of flavors and fragrances, finished perfume compositions in a very big way. However, small and medium enterprises (SMES) have been also actively engaged in promotion of the particular industry in a significant way. Thanks to important contribution made by all sectors of industry it is expected that flavor and fragrance industry encompassing essential oils and finished perfumes will make very important strides in this industry.

GLOSSARY

1. Essential Oils

A volatile oil derived from the leaves, stem, flower or twigs of plants, and usually carrying the odor or flavor of the plant. Chemically they are mostly terpenes (hydrocarbons), but many other types occur.

They have pungent taste and odor, usually nearly colorless when fresh but becoming darker and thick on exposure to air, soluble in alcohol, CS_2, CCl_4, chloroform, petroleum ether and fatty oils.

Obtained by steam distillation of odorous components of plants, also by solvent extraction.

2. Fixed Oils

A name give to a vegetable oil obtained from plants that, in contradistinction to essential oils, are fatty, dense and non-volatile (boiling points 200-300°C) such as olive or sweet almond oil.

3. Concrete

A concentrate, waxy, solid or semi-solid perfume material prepared from previously live plant matter usually using a hydrocarbon type of solvent such as benzene, toluene, hexane, petroleum ether etc.

4. Pomades

Consist of fats that contain fragrance substances and are produced by the hot or cold effleurage of flowers.

5. Absolutes

Substances that are prepared by extraction of concretes using dry ethanol or any similar solvent. Compounds that precipitate on cooling are then removed by filtration. After evaporation of the solvent like ethanol a wax-free residue called and absolute is left behind.

6. Herb

Any seed-bearing plant which does not have a woody steam and dies down to the ground after flowering. Basil, bay leaf, mint, thyme etc. are typical herbs.

7. Shrub

A small woody plant much smaller than a tree and has several main stems arising at or near the ground.

8. Aroma Therapy

It is the art and science of using aromatic plants primarily in the form of essential oils for strengthening and maintaining human wellbeing or for removing diseases when physical, emotional, mental or spiritual imbalance has occurred.

9. Carrier Oils

These oils, many of them essential, some of them even as fixed oils, predominantly are used as blended oils in aroma therapy, to dilute the concentration of rich and powerful essential oils, are rich in vitamins and minerals. Recommended for use as massage oils or otherwise, in aroma therapy.

10. Enfleurage

Extraction of odoriferous components of flowers by means of fats or mixtures of fats and tallow, the process being carried out at room temperature to avoid decomposition of the desired perfumes. The latter are separated from the fat by washing with alcohols.

11. Herbalism

The art and science of use of botanicals, more particularly herbs, for medicinal, therapeutic and culinary uses has been broadly classified as herbalism. Basil, bay leaf, bergamot, dill, fennel, parsley, spearmint, thyme etc. have been among the prominent botanicals used for often for herbalism.

12. Balsam

A resinous semi-solid mass or viscous liquid exuded from a plant, which can be either a pathological or a physiological product. A true balsam is characterized by its high content of benzoic acid, benzoates, cinnamic acid or cinnamates.

13. Balsamic
A soothing medicine or application having the qualities of a balsam.

14. Disinfectant
That prevents and also combats the spread of germs.

15. Gums
True gums are little used in perfumery, being virtually odorless. However, the term 'gum' is often applied to resins, especially with relation to turpentines, as in the Australian 'gum tree', strictly speaking gums are natural or synthetic water-soluble materials such as gum Arabic.

16. Narcotic
A substance that induces sleep, intoxicating or poisonous in large doses.

17. Oleoresin
A natural resinous exudation from plants, or an aromatic liquid preparation, extracted from botanical matter using solvents. They consist almost entirely of a mixture of essential oil and resin.

18. Resinoids
A perfumery material prepared from natural resinous matter, such as balsams, gum resins, etc., by extraction with a hydrocarbon type of solvent.

19. Phytotherapy
The treatment of disease by plants; herbal medicine.

20. Olfaction
The sense of smell.

21. Tannin
A substance that has an astringent action, and helps seal the tissue.

22. Tuber
A swollen part of an underground stem of one year's duration, capable of new growth.

23. Volatile

Unstable, evaporates easily, as in 'volatile oil'.

24. Stimulant

An agent that quickens, the physiological function of the body.

25. Synergy

Agents working together harmoniously, assisting in co-ordination in the action of muscles, organs or substances such as drugs.

References

1) Ullman's Encyclopedia of Industrial Chemistry, 5th edition, Germany, 1996, Perfumes, Vol. A19, pp 171-176.

2) Lewis, J.R., In Hawley's Condensed Chemical Dictionary, 13th edition; Galgotia Publications Pvt. Ltd., New Delhi, 1998.

3) Shreve R. Norris, Brink Joseph A. (Jr), Chemical Process Industries, 4th edition, (McGraw-Hill International Book Company), Tosho Printing Co. Ltd., Tokyo Japan, 1977.

4) Lawless Julia, in The Illustrated Encyclopedia of Essential Oils, Barnes & Noble Books, New York, 1995, pp 50-53.

5) Oils, Essential, in Krik-Othmer Encyclopedia of Chemical Technology, 4th edition, Wiley-Inter Science Publications, New York, 1994, Vol. 16, pp 307-314.

6) Mahindru, S.N., Indian Plant Perfumes, Metropolitan Book Co. Pvt. Ltd., New Delhi, 1992, pp 13-21.

7) Mukhopadhyay, A.K., Industrial Chemical Cresols and Downstream Derivatives, Marcel Dekker, New York, 2005, pp. 162-164.

8) Some Essential Oils - Materials and Technology, Vol.V – J.H. de Bussy, Amsterdam, The Netherlands, 1972, pp 1-26.

9) Bhargava Kush, Aroma Therapy, in Handbook of Alternative Medicine, Love Kush Publication, Surat, Gujarat, India, September 2003, pp 41-64.

10) Ullman's Encyclopedia of Industrial Chemistry, 5th edition, Germany, 1996, Flavors and Fragrances, Vol. A11, pp 214-245.

11) Dodt K. Collean, The Essential Oils Book, Storey Books, Schoolhouse Road, Pownal, Vermont 05261, USA, 1996, pp 40 -60.

12) Merck Index, 13th edition, Merck & Co., Inc., New Jersey, USA, 2001.

13) Shah, C.S., Quadry, J.S., A Text Book of Pharmacognosy, Volatile Oils, 9th edition, Shah Prakashan, R.S., Ahmedabad, India, 199293, pp 177-250.

14) Organic Chemicals from Natural Sources, Sevak Publications, Wadala, Bombay 400 031, 1987, Essential Oils, pp 84-105.

15) Mukhopadhyay A.K., Flavors and Fragrances from Plant Sources, Proceedings on Botanical Products Seminar and Expo-2005, 5-7 February 2005, Jaipur, Rajasthan, India, pp 273-280.

16) National Convention on Essential Oils & Isolates, December 2-3, 2005, Small Industries Service Institute, Mumbai, India.

17) Lindley David & Moore T. Harvey Webster's New World Dictionary of Science', Helicon Publishing Limited, 1998, New York, USA.

18) Satoskar, K.G., Constraints in Isolation of Aroma Chemicals from Essential Oils, National Convention of Essential Oils & Isolates for SMES, December 2-3, 2005, Mumbai, pp 70-72.

19) Harlalka Ramakant, Oils from Forestry Sector, National Convention of Essential Oils & Isolates for SMES, Dec. 2-3, 2005, Mumbai, pp 64-68.

20) Walavalkar, Shashi - Role of Essential Oils in Perfumery Blending, National Convention on Essential Oils & Allied Industries for SMES, October 1-2, 2004, Mumbai.

21) Perfumes, Kirk-Othmer Encyclopedia of Chemical Technology, 4th edition, Wiley Inter-Science Publications, New York, 1994, Vol. 18, pp 171-181.

22) Mookherjee, B.D., Trenkle, R.W., Wilson, R.A., Pure Applied Chemistry, 62, 1357 (1990).

23) Bedoukian, P.Z., Perfumery and Flavoring Synthetics, Allured Publishing Corp., Wheaton, Illinois, USA, 1986.